Advance

WIND POWER Basics

Renewable energy systems are a future imperative that
Dan Chiras' approach grounds firmly in the present.
This thorough, practical work details small-scale and
commercial wind power in terms accessible to any level of
interest. Akin to the exuberant expertise of Julia Child, this
book could be called "Mastering the Art of Wind Power."

Sp

book had been available when I first started my business. Now
that I'm training small wind installers, I will recommend this
book to all my students. This book is a wonderful source of
basic wind info and a must have for small wind newbies!

— Roy Butler, NABCEP Certified Solar PV Installer®,
NYSERDA eligible PV & wind installer,
PA Sunshine Program Approved PV Installer

In *Wind Power Basics*, Dan has again shown his true
brilliance gained from years of passionate commitment to,
and teaching of, sustainability. He takes on a vast subject,
understanding it thoroughly through his own study and
experience, and then presents it in a concise, complete and
easy-to-understand manner. He has covered everything one
needs to know about small wind systems in one volume.
For anyone contemplating a wind system, or simply wanting
to understand them, this book is the definitive guide.

— James Plagmann, AIA + LEED AP — Green Architect

I've followed the pioneering work of Dan Chiras for many years, watching him emerge as the go-to guy for energy self-reliance. Through meticulous research and many hands-on projects, Chiras has acquired essential, how-to knowledge about small-scale renewable energy. In *Wind Power Basics*, he shares what he knows in a very accessible way. Get this book onto your kitchen table and spend pleasant time with it!

— David Wann, coauthor of *Affluenza* and author of *Simple Prosperity*

Dan Chiras is as reliable as a Swiss watch — once again he's created a text that's as accessible as it is informative. I look forward to recommending this book to all my clients who are exploring wind energy.

— Ann Edminster, author, *Energy Free: Homes for a Small Planet*

WIND
POWER
Basics

DAN CHIRAS

With Mick Sagrillo and Ian Woofenden
and technical advisors
Robert Aram, PE and Jim Green, PE
Illustrations by Anil Rao, Ph.D.

NEW SOCIETY PUBLISHERS

Cataloging in Publication Data

A catalog record for this publication is available from
the National Library of Canada.

Cover design by Diane McIntosh.
Cover images © iStock: background leaf: alphacat;
wind turbine: Arpad Benedek; electrical cord: Diane Labombarbe
Illustrations by Anil Rao, Ph.D.

Printed in Canada by Friesens.First printing April 2010

Paperback ISBN: 978-0-86571-617-9

Inquiries regarding requests to reprint all or part of *Wind Power Basics*
should be addressed to New Society Publishers at the address below.
To order directly from the publishers, please call toll-free (North America)
1-800-567-6772, or order online at
www.newsociety.com

Any other inquiries can be directed by mail to:

New Society Publishers
P.O. Box 189, Gabriola Island, BC V0R 1X0, Canada
(250) 247-9737

New Society Publishers' mission is to publish books that contribute in funda-
mental ways to building an ecologically sustainable and just society, and to do so
with the least possible impact on the environment, in a manner that models this
vision. We are committed to doing this not just through education, but through
action. This book is one step toward ending global deforestation and climate
change. It is printed on Forest Stewardship Council-certified acid-free paper that
is **100% post-consumer recycled** (100% old growth forest-free), processed chlorine free, and printed with vegetable-based, low-VOC inks, with covers produced
using FSC-certified stock. New Society also works to reduce its carbon footprint, and purchases carbon offsets based on an annual audit to ensure a carbon
neutral footprint. For further information, or to browse our full list of books
and purchase securely, visit our website at: www.newsociety.com

NEW SOCIETY PUBLISHERS

Mixed Sources
Cert no. SW-COC-001271
© 1996 FSC
FSC

Contents

Dedication

To my dear friend, Ed Evans,
whose dedication to teaching is unrivaled and
whose friendship over the years has been a
gift of immeasurable value to me.

Preface

My work on this book and my previous book, *Power from the Wind*, started in the summer of 2006. I'd just published *The Homeowner's Guide to Renewable Energy*, a book that introduces readers to the many renewable energy options they can tap into to power their homes and businesses.

I was working with a client in northern Michigan, helping her design a super-efficient passive solar home for a small organic farm and ecological learning center. She asked if I'd design a wind energy system to produce electricity to pump water to irrigate her gardens. I answered yes without thinking. What could be so hard about that?

When I arrived home, I realized that I needed to learn a lot more about small wind energy systems and site analysis to provide meaningful advice, so I began to read everything I could on small wind energy systems. Soon after my studies began, I became enthralled with this technology. As my knowledge grew, I felt the fervent didactic impulse emerge once again. The impetus for all my books has been a desire to share the new and exciting knowledge I've acquired. By sharing this knowledge, I like to think I am making a contribution to individuals who want to learn about wind but don't have the time or inclination to wade through a hundred articles and books on the subject. As my knowledge grew, however, I realized that I needed help. I needed a knowledgeable expert or two in small wind energy to provide advice and guidance, correct mistakes, and provide additional information. While attending the Midwest Renewable Energy Association's annual energy fair, I asked Mick Sagrillo if he'd help. Mick's the guru of small wind. He's one of the most knowledgeable small wind experts in the world. Mick Sagrillo has been in the small wind industry since 1981. He has taught numerous workshops on small wind energy systems through the Midwest Renewable Energy Association and Solar Energy International. He writes a column for *Windletter*, the American Wind Energy Association's newsletter, and *Solar Today*. He has also published numerous articles in *Home Power* magazine and has consulted on Wisconsin's Focus on Energy's small wind program. Despite a hectic schedule, Mick agreed to assist me. He served as the technical advisor to me on my wind energy book, offering invaluable information and advice throughout.

A year or so later, I approached New Society Publishers with the idea of publishing a series of books on renewable energy technologies for the home and office. They liked the idea, and we were off and running.

It's then that I shifted my education into high gear. I took numerous workshops on small wind energy. Most important were hands-on workshops where we installed entire wind systems. In 2007, I became a certified wind site assessor. In 2008, I built my own wind turbine.

In 2007, I recruited another wind expert, *Home Power's* Ian Woofenden. Like Mick, Ian teaches installation workshops and writes articles for *Home Power* magazine on wind energy. He's also a gold mine of information, as he's been living off-grid on wind and solar energy for years.

In the summer of 2007, Jim Green, the National Renewable Energy Laboratory's small wind expert offered his assistance. He volunteered to read the manuscript and to help ensure the book's accuracy.

A few months later, I added Robert Aram, an electrical engineer, to the team. I'd worked with Bob in several wind energy workshops. His vast knowledge, experience, and extraordinary ability to explain complex subjects in a clear way and his knowledge of physics and engineering proved extremely valuable to me. Bob read the book to help ensure accuracy.

My advisors offered valuable comments that helped me produce *Power from the Wind*, published in 2009, and now Wind Power Basics, the condensed version you hold in your hands. I am deeply indebted to them for their comments and corrections and am extremely honored to have worked with such amazing people. A world of thanks to them and to all the others whose work I relied on when writing this book.

A world of thanks also to my dear friends at New Society Publishers, Chris and Judith Plant, who have, over the years, been an absolute delight to work for. Thanks for taking this project on and for believing in me, and thanks for their unwavering dedication to creating a sustainable society. Thanks to the staff at New Society as well: Ingrid Witvoet who shepherded this book through production; Greg Green, who, as always, did a smashing job of designing and laying out the book; and my astute and congenial copyeditor, Linda Glass, for helping make this a better book. I'd also like to thank a new member of my team, Dr. Anil Rao, a professor of biology, who illustrated this book. He's a valuable part of this book's success.

I'd also like to thank my family. A world of thanks to my partner Linda who has listened patiently to my many discussions of wind turbines and wind site assessments. Thanks, too, to my sons, whose lives continue to grace mine.

— Dan Chiras, Evergreen, Colorado

Books for Wiser Living
recommended by *Mother Earth News*

Today, more than ever before, our society is seeking ways to live more conscientiously. To help bring you the very best inspiration and information about greener, more sustainable lifestyles, *Mother Earth News* is recommending select New Society Publishers' books to its readers. For more than 30 years, *Mother Earth* has been North America's "Original Guide to Living Wisely," creating books and magazines for people with a passion for self-reliance and a desire to live in harmony with nature. Across the countryside and in our cities, New Society Publishers and *Mother Earth* are leading the way to a wiser, more sustainable world.

CHAPTER 1

INTRODUCTION TO SMALL-SCALE WIND ENERGY

Humans have harvested energy from the wind for centuries. Harnessed by the Europeans as early as 900 years ago, wind was used to grind grain and manufacture goods. Wind powered ships that helped open up new territories, spurring international trade. In North America, wind energy has been used since the late 1800s. Over the years, tens of thousands of farms in the Great Plains relied on wind pump water for livestock and domestic uses — some still do.

Windmills began to emerge in the 1860s in rural America. By 1890, there were over 100 manufacturers of water-pumping windmills (Figure 1.1). All told over 8 million were installed in this country. Many of these water-pumping windmills have been restored and are still operating today with minimal maintenance.

Windmill vs. Wind Turbine

A windmill is a machine that converts the energy of the wind into other, more useful forms like mechanical energy. Early windmills were designed to grind grain and pump water. Later on, windmills were designed to generate electricity. Electricity-generating windmills are commonly referred to as wind turbines or wind generators. Water-pumping windmills are generally referred to as such or simply as windmills.

Wind energy was also extremely important to railroads in the West. Windmills were often used to fill water tanks along tracks to supply the steam engines of locomotives.

In the 1920s through the early 1950s, many Plains farmers also installed wind turbines to generate electricity. The turbines powered lights and all their appliances, many of which were ordered from the Sears catalog — including electric toasters, washing machines and radios. Radios were particularly important, as they allowed farmers and their families to keep in touch with the world.

Unfortunately, the use of water-pumping and small wind-powered electric generators began to decline in the United States in the late 1930s. Their demise was due in large part to America's ambitious Rural Electrification Program. This program, which began in 1937, was designed to provide electricity to rural America. As electric service became available, wind-electric generators were mothballed.

Fig. 1.1: *The Old and the New. Water-pumping windmills like the one in the foreground were once common in the West and Midwest. The technology hasn't changed in 100 years. In the distance is a modern commercial wind turbine that generates electricity to power cities and towns.*

DAN CHIRAS

In fact, local power companies required farmers to dismantle their wind generators as a condition for providing service via the ever-growing electrical grid. The electrical grid, or simply the grid, is the extensive network of high-voltage electrical transmission lines that crisscross nations, delivering electricity generated at centralized power plants to cities, towns and rural customers. A key advantage of the grid was its ability to provide virtually unlimited amounts of electricity to customers.

Unfortunately, rural electrification drove virtually all of the manufacturers of windmills and wind-electric generators out of business by the early 1950s. However, in the mid-1970s, wind energy made a resurgence as a result of intense interest in energy self-sufficiency in the United States, stimulated principally by back-to-back oil crises in the 1970s that resulted in skyrocketing oil prices and a period of crippling inflation. Generous federal incentives for small wind turbines, incentives from state governments, and changes in US law that required utilities to buy excess electricity from small renewable energy generators helped stimulate the comeback.

Soon thereafter, however, wind energy took a nosedive. Conservation and energy efficiency measures in the United States and new, more reliable sources of oil drove the price of oil and gasoline down. Federal and state renewable energy tax incentives disappeared as a result of a precipitous decline in America's concern for energy independence. As a result, all but a handful of the small wind turbine manufacturers went out of business.

In the 1990s, commercial and residential wind energy staged another comeback as a result of many factors, among them rising oil prices, global awareness of the decline in world oil production, an increase in the cost of natural gas, and growing concern for global climate change and its impacts.

Because of these factors, many believe that this time around, wind energy is here to stay. Much to the delight of renewable energy advocates, large commercial wind farms have begun to appear in numerous countries, most notably the United States, Germany, Spain

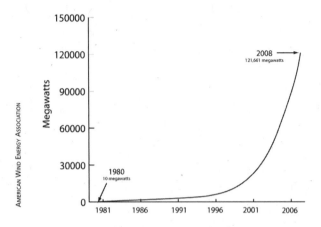

AMERICAN WIND ENERGY ASSOCIATION

Fig. 1.2: *Global Wind Energy Capacity. This graph shows the installed global capacity (in megawatts) of commercial wind turbines.*

and Denmark. These facilities produce huge amounts of electricity and are changing the way the world meets its energy needs. Today, wind-generated electricity is the fastest growing source of energy in the world (Figure 1.2).

Although commercial wind farms are responsible for most of the growth in the wind industry, smaller residential-scale wind machines are also emerging in rural areas, supplying electricity to homes, small businesses, farms, ranches and schools (Figure 1.3). Most of the small-scale wind turbines "feed" the excess electricity they produce back onto the electrical grid.

World Wind Energy Resources

Wind energy is clearly on the rise and could become a major source of electricity in years to come because wind is widely available and often abundant in many parts of the world. Significant resources are found on every continent. Tapping into the world's windiest locations could theoretically provide 13 times more electricity than is currently produced worldwide, according to the Worldwatch Institute, a Washington, DC-based nonprofit organization.

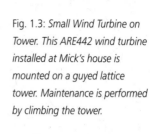

Fig. 1.3: *Small Wind Turbine on Tower. This ARE442 wind turbine installed at Mick's house is mounted on a guyed lattice tower. Maintenance is performed by climbing the tower.*

DAN CHIRAS

Rated Power in Watts or Kilowatts

Wind turbines are commonly described in terms of rated power, also known as rated output or rated capacity. Rated power is the instantaneous output of the turbine (measured in watts) at a certain wind speed (called the rated speed) at a standard temperature and altitude. The rated power of small wind turbines falls in the range of 1,000 to 100,000 watts. One thousand watts is one kilowatt (kW). Large wind turbines include all of those turbines over 100 kilowatts. Most larger turbines, however, are rated at one megawatt or higher. A megawatt is a million watts or 1,000 kilowatts. It is important to note that wind turbines do not produce their rated power all of the time, only when they're running at their rated wind speed. As noted in Chapter 5, while rated power is commonly used when describing wind turbines, it is one of the least useful and most misleading of all parameters by which to judge a wind generator.

In North America, wind is abundant much of the year in the Great Plains and in many northern states. It is also a year-round source of energy along the Pacific and Atlantic coasts and the shores of the Great Lakes. Tapping into the windiest locations in the United States, for example, in North and South Dakota could produce enough electricity to supply *all* of the nation's electrical needs. Proponents of wind energy estimate that wind could eventually provide at least 20 to 30 percent of the electricity consumed in the United States and other countries.

Proponents of renewable energy envision a future powered by wind and a host of other clean, affordable renewable energy resources, among them solar energy, biomass, geothermal energy, tidal energy, wave energy and ocean currents (Figure 1.4).

The Pros and Cons of Wind Energy

Wind is a seemingly ideal fuel source that could ease many of the world's most pressing problems. Like all energy sources, small wind power has its advantages and disadvantages. Let's look at the downsides of small wind systems first.

Fig. 1.4: *Solar Array. In a renewable energy future, large solar electric installations like this one will supplement electricity produced by other renewable resources, including wind, hydropower and biomass.*

Disadvantages of Wind Energy

Small wind's disadvantages are few and often grossly exaggerated or only perceived problems. They include wind's variability, bird mortality, aesthetics, property values and unwanted sound. Some people are concerned about wind being more site specific than solar electricity. There's even concern about ice falling from turbines after ice storms and interference with radio and televisions signals.

Variability and Reliability of the Wind

Perhaps the most significant "problem" with small wind is that the wind does not blow 100 percent of the time in most locations. Wind is a variable resource, to be sure. It's not available 24 hours a day like coal or oil. In fact, a wind turbine may operate for four days in a row, producing a significant amount of electricity, then sit idle for two days — or a week.

Wind resources vary seasonally, too. In most locations, winds are typically strongest in the fall, winter and early spring, but decline during the summer. Wind even varies during the course of a day. Winds may blow in the morning, die down for a few hours, then pick up later in the afternoon and blow throughout the night.

Even though wind is a variable resource, it is not unreliable. Just like solar energy, you can count on a certain amount of wind at a given location during the year. With smart planning and careful design, you can design a wind system to meet your electrical needs.

Wind's variable nature can be managed to our benefit by installing batteries to store surplus electricity in off-grid systems. The stored electricity can power a home or office when the winds fail to blow.

Surplus electricity can also be stored on the electrical grid in many systems. Thus, when a wind-electric system is producing more power than a home or business is using, the excess is fed onto the grid. In times of shortfall, electricity is drawn from the grid.

Wind's variable nature can also be offset by coupling small wind systems with other renewable energy sources, for example, solar-electric systems. Such systems are referred to as hybrid systems. Solar-electric systems (or photovoltaic [PV] systems) generate electricity when sunlight strikes solar cells in solar modules. Hybrid systems can be sized to provide a steady year-round supply of electricity. Residential wind-generated electricity can also be supplemented by small gas or diesel generators.

Bird Mortality

One perceived problem with wind power is bird mortality. Unfortunately, this issue has been blown way out of proportion. Although a bird may occasionally perish in the spinning blades of a residential wind machine, this is an extremely rare occurrence. Ian is aware of only one instance of a bird kill, when a hawk flew into a small wind turbine. "Because of their relatively smaller blades and short tower heights, home-sized wind machines are considered too small and too dispersed to present a threat to birds," notes Mick Sagrillo in his article, "Wind Turbines and Birds," published by Focus on Energy, Wisconsin's renewable energy program.

The only documented bird mortality of any significance occurs at large commercial-scale wind turbines — but even then, the number of deaths is relatively small. Commercial wind turbines kill an estimated 50,000 birds per year. While this may sound like a lot, this number pales in comparison to other lethal forces, among them domestic cats, automobiles, windows in buildings, and communication towers. All in all, cats are probably the most lethal "force" that birds encounter. Scientists estimate that our beloved cats kill about 270 million birds a year nationwide — though the number is very likely much higher.

Aesthetics

Although many people view small wind turbines as things of great beauty, others contend that they detract from natural beauty. Ironically, those who find wind turbines to be unsightly often ignore the great many forms of visual blight in the landscape, among them cell phone towers, water towers, electric transmission lines, radio towers and billboards. To be fair, there are differences between a wind tower and common sources of visual pollution. For one, a wind turbine's spinning blades call attention to these machines. Another is that we've grown used to the ubiquitous electric lines and radio towers. As a result, people often fail to see them anymore. Given the opportunity to oppose a structure in their "viewshed" (for example, at a public hearing that may be required for permission to install a residential wind system) neighbors will often speak up in opposition. If you need to apply for permission to install a turbine on a tall tower, you may encounter this problem. We'll talk about ways to address this in the last chapter.

Proximity to Homes and Property Values

Critics raise legitimate concerns when it comes to the placement of wind machines near their property. Although most of the issues over proximity have been raised by individuals and groups that oppose large commercial wind farms, residential systems can also cause a stir among neighbors. Some may be concerned about aesthetics. Others may worry about safety.

To avoid problems, we recommend installing machines in locations out of sight and hearing of neighbors. Safety concerns are typically related to tower collapse — an extremely rare event that is always the result of bad design and improper installation. Even though homeowner's insurance should cover damage to individuals and property, it is best to place a wind turbine and tower well away from your neighbors' property lines.

Unwanted Sound

Opponents of wind energy and apprehensive neighbors sometimes voice concerns about unwanted sound, a.k.a. noise, from residential wind machines. Small wind turbines do produce sound, and as the wind speed increases, sound output increases. Sound is produced primarily by the spinning blades and alternators. The faster a turbine spins, the more sound it produces.

You can reduce unwanted sound by selecting a quieter, low-rpm wind turbine rather than a louder, high-rpm wind turbine. If you are concerned about sound, make this a high priority as you shop for a turbine and let your neighbors know you are sensitive to this issue.

Wind turbines have governing mechanisms, systems that slow down the machines when winds get too strong to protect them from damage. Different governing systems result in different sound levels. (We'll discuss this topic in Chapter 5.) When researching your options, we recommend that you listen to the turbines you're considering buying in a variety of wind conditions, including those that require governing.

To reduce sound at ground level, be sure to mount your turbine on a tall tower. Suitable tower heights, which we'll discuss later, are usually 80 to 120 feet. A residential wind turbine mounted high on a tower catches the smoother and stronger — and hence most productive — winds. This strategy also helps reduce sound levels on the ground because sound dissipates quickly over distance.

Residential (and commercial) wind machines are also much quieter than many people suspect because the sounds they make are partially drowned out by ambient sounds on windy days. Rustling leaves and wind blowing past one's ears often drown out much of the sound produced by a residential wind turbine.

Sound is measured in two ways — by loudness and frequency. Loudness is measured in decibels (dB). Frequency is the pitch. A low note sounded on a guitar has a low frequency or pitch. A high note has a high frequency. The average background noise in a house is about 50 dB. Nearby trees on a breezy day measure about 55 to 60 dB. Most of today's residential wind turbines perform very near ambient levels over most of their operating range.

Even though the intensity of sound produced by a wind generator may be the same as ambient sound, the frequency may differ. As a result, wind turbine sounds may be distinguishable from ambient noises, even though they are not louder. You'll hear a swooshing sound. In other words, while the sound of a wind turbine can be picked out of surrounding noise if a conscious effort is made to hear it, home-sized wind turbines are not the noisy contraptions that some people make them out to be.

Site Specific

Yet another criticism of small wind is that it is more site specific — or restricted — than solar energy.

To understand what this means, we begin by pointing out that there are good solar areas and good wind areas. In a good solar region, most people with a good southern exposure can access the same amount of sun. In a windy area, however, hills and valleys or stands of trees can dramatically reduce the amount of wind that blows across a piece of property. Therefore, even if you live in an area with sufficient winds, you may be unable to tap into the wind's energy because of topography or nearby forests or stands of tall trees. That's what critics mean when they say that wind energy is more site specific than solar.

That said, we should point out that solar resources also vary. If you live in a forest in a sunny location, for example, you'll have a lot less solar energy than a nearby neighbor whose home is in a field. In addition, homeowners can access the wind at less-than-optimum sites by installing turbines on tall towers. Tall towers help you overcome topographical and other barriers.

Ice Throw

Like trees and power lines, wind turbines can ice up under certain conditions. Ice falling off the blades is known as ice throw, and is a concern that may arise during zoning hearings on residential wind turbines.

While ice builds up on blades and wind turbine towers during ice storms, it is typically deposited in very thin sheets. When the blades are warmed by sunlight, the ice tends to break up into small pieces, not huge dangerous chunks, and drop to the ground.

Ice buildup on the blades of a wind turbine dramatically reduces the speed at which a turbine can spin. It's a little like trying to drive a car with four flat tires. As a result, ice is not thrown from a turbine, it falls around the base of the tower — just as it does from trees and power lines.

Any prudent person would stay away from the tower base when ice is shed from the blades, as they would from trees or power lines covered with ice warming in the sun. Ice-laden trees are also considerably more dangerous, as branches can and often do break and fall to the ground, damaging power lines, cars and houses. Entire trees can topple as a result of ice buildup.

On the rare occasion that ice builds up on a wind turbine, experienced wind turbine operators shut down their machines until the Sun or warmer temperatures melt the ice since they cannot generate electricity spinning at such low revolutions per minute anyway.

Interference with Telecommunications

Some opponents of wind energy raise the issue of interference with telecommunications signals. This is simply not a problem. Turbines for homes and small businesses have small blades that do not interfere with such signals. Moreover, the blades of modern wind turbines are made out of materials that are "transparent" to telecommunications signals. As a result, small wind turbines are often installed to power remote telecommunications sites. Telecommunication equipment wouldn't be installed in such locations if there were a problem with interference.

The Advantages of Wind Energy

Although residential wind turbines and their energy source, the wind, have a few downsides, wind energy is an abundant and renewable resource. We won't run out of wind for the foreseeable future, unlike oil and natural gas.

Small-scale wind energy could also help decrease our reliance on declining and costly supplies of oil — if electricity generated by wind is used to power electric or plug-in electric hybrid cars and trucks, displacing gasoline, which is refined from oil.

Wind energy can also play a meaningful role in offsetting declining US natural gas supplies. In the United States, approximately 18 percent of all electricity is currently generated by natural gas, according to the US Department of Energy. As supplies decline, wind could help ease the crunch, supplying a growing percentage of our nation's electricity.

Wind could even eventually reduce our dependence on nuclear power as well. In the United States, nuclear power plants generate about 20 percent of the nation's electricity. Although wind energy does have its impacts, it is a relatively benign technology compared to conventional sources of electricity. It could help all countries create cleaner and safer energy at a fraction of the environmental cost of conventional electrical energy production. Wind energy can help nations reduce global warming and devastating changes in our climate. Wind can also help homeowners and businesses do their part in solving other costly environmental problems such as acid rain.

Another benefit of wind energy is that, unlike oil, coal and nuclear energy, the wind is not owned by major energy companies or controlled by foreign nations. An increasing reliance on wind energy could therefore ease international political tension. Reducing our reliance on Middle Eastern oil could reduce costly military operations aimed, in part, at stabilizing a region where the largest oil reserves reside.

Wind is also a free resource. The cost of wind is not subject to price increases. A wind- and solar-powered future might be one subject to less inflation. This is not to say that wind energy will be free of price increases. While the fuel itself (the wind) is free, the price of wind generators is likely to increase. That's because it takes energy to extract and process minerals to make the steel and copper needed for wind turbines and towers. It also takes energy to make turbines and towers and ship and install them. As the price of conventional fuels and raw materials increases, the cost of wind energy also will go up.

Yet another advantage of wind-generated electricity is that it uses existing infrastructure, the electrical grid, and existing technologies. A transition to wind energy could occur fairly seamlessly.

Thanks to generous tax credits and other financial incentives, individuals in rural areas with good wind resources can meet all or part of their energy needs at rates that are often competitive with conventional sources. In remote locations, wind or wind and solar electric hybrid systems can be cheaper than conventional power,

Fig. 1.5: *Plug-In Hybrid. Electric cars and plug-in hybrids like the one shown here are the most promising automobile technologies on the horizon. They could be powered by electricity from the Sun and wind.*

which requires the installation of costly electric lines that transport electricity from power plants to end users.

The Purpose of this Book

This book's principal focus is on small wind-electric systems — those with rated output ranges from 1 kilowatt to 100 kilowatts. Most of the turbines we'll be discussing fall in the 1- to 20-kilowatt (kW) range. The blades of small wind turbines (1 to 100 kW) range from 4 feet to 32 feet in length. Small-scale wind systems serve a variety of purposes. The smallest units are generally sufficient to power cabins and cottages; larger small wind turbines power homes and small businesses as well as schools, farms, ranches, small manufacturing plants, and public facilities.

This book is written for individuals who want a succinct introduction to small-scale wind systems that doesn't require a degree in physics or engineering. My goal was to create a user-friendly book that teaches readers the basics of wind energy and wind energy systems. This book is *not* an installation manual, but it will help you learn about wind if you want to become a wind energy installer or install a wind turbine and tower on your property. It will also help you determine if wind energy is right for you, what your options are, and how much it is going to cost.

When you are done with this book, you should have a good knowledge of the key components of wind energy systems. This book will help you when shopping for a wind system or an installer. You'll also learn about maintenance requirements.

If you choose to hire a professional wind energy expert to install a system — a route we highly recommend — you'll be thankful you've read this book. The more you know, the more input you will have into your system design, components, siting and installation — and the more likely that you'll be happy with your purchase.

This book should help readers develop realistic expectations. Wind energy systems, for instance, require annual inspection and maintenance — climbing or lowering a tower to access the wind turbine to check for loose fasteners and blade damage and, much less commonly, an occasional part replacement. If you are not up for it or don't want to pay someone to climb or lower your tower once or twice a year to check things out, you may want to invest in a solar electric system instead.

Organization of this Book

After this brief introduction, we turn our attention to the wind itself, the driving force in a wind energy system. In Chapter 2, you will learn how winds are generated and explore the factors that influence wind flows in your area. We will also explore the factors that affect energy production by a residential wind turbine and why it is important to mount a wind machine on a tall tower.

In Chapter 3, we'll explore small wind energy systems. You'll learn the three types of residential wind energy systems: (1) off-grid, (2) batteryless grid-tie, and (3) grid-connected with battery backup. You'll learn about the basic components of each one and hybrid wind systems.

In Chapter 4, you will learn how to assess your electrical energy needs and how to determine if your site has enough wind to make a wind system worthwhile. You'll learn why energy conservation and energy efficiency will save you a lot of money on a wind energy system. You will also learn ways to evaluate the economics of a wind system answering the question: Does a wind system make sense from an economic perspective?

Chapter 5 will introduce you to wind turbines — what types are available and how they work. We'll also give you shopping tips — what to look for when buying a wind turbine. We'll spend a little time discussing building your own wind generator.

Chapter 6 describes basic tower options, the pros and cons of each one, and how they are installed.

In Chapter 7, we'll study storage batteries and charge controllers, two key components of off-grid wind systems.

Chapter 8 addresses another key component of all wind energy systems, the inverter. You will learn how inverters work, what functions they perform, and what to look for when shopping for one.

In Chapter 9, we'll give a brief overview of wind energy system maintenance. We'll also explore a range of issues such as homeowner's insurance, financing renewable energy systems, building permits, electrical permits and zoning. This books ends with a resource guide that lists books, magazines, organizations, small wind turbine manufacturers and wind turbine tower kits.

CHAPTER 2

WIND AND WIND ENERGY

As you learned in the last chapter, wind is a clean, abundant, and renewable energy resource that can be tapped to produce electricity. This chapter explores how wind is generated and introduces you to two types of wind — local and global. We'll also explore ways local topography affects wind, introducing you to two key concepts: ground drag and turbulence. This information provides the practical knowledge you will need to select the best site for a wind turbine and the optimum tower height.

What is Wind?

Wind is air in horizontal motion across the Earth's surface. All winds are produced by differences in air pressure between two regions. Differences in pressure result from differential heating of the surface of the Earth. Heating, of course, is caused by sunlight striking the Earth's surface.

Like most other forms of energy in use today, even coal, oil and natural gas, wind is a product of sunlight — solar energy. Some wind advocates, refer to wind as "the other solar energy" or "second-hand solar energy." Let's begin by looking at two types of local winds: (1) offshore and onshore winds and (2) mountain-valley breezes.

Offshore and Onshore Winds

Offshore and onshore winds are generated along the shores of large lakes, such as the Great Lakes of North America, and along the coastlines of the world's oceans. Offshore and onshore winds blow regularly, nearly every day of the year. They are produced by the differential heating of land and water, caused by solar energy.

Here's how this happens: As shown in Figure 2.1a, sunlight shining on the Earth's surface heats the land and water simultaneously. As the water and adjoining land begin to warm, they radiate some of the heat (infrared radiation) into the atmosphere. This heat, in turn, warms the air above them. When air is heated it expands, and as it expands it becomes less dense and rises. The upward movement of air is called a thermal or updraft.

Although water and land both heat up when warmed by the sun, land masses warm more rapidly than neighboring bodies of water. Because air over land heats up more quickly than air over water, air pressure over land is lower than over neighboring surface waters. As warm air rises over land, cooler, high pressure air moves in to fill the void, resulting in a steady breeze known as onshore wind.

At night, the winds blow in the opposite direction — from land to water — as illustrated in Figure 2.1b. These are known as offshore breezes or offshore winds.

Like onshore winds that occur during the day, offshore winds are created by differences in air pressure between the air over land and neighboring water bodies. Here's what happens: after sunset, the land and the ocean both begin to cool. Land, however, cools more rapidly than water. Because the water cools more slowly, air above it is warmer. Warm air expands and rises. Cooler high pressure air

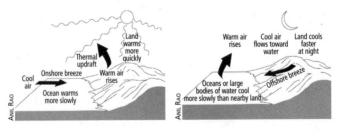

Fig. 2.1a and 2.1b: *Onshore and offshore breezes. Onshore (a) and offshore (b) breezes occur along the coastlines of major lakes and oceans.*

flows from the land to the water at night (Figure 2.1b). The result is an offshore breeze: steady winds that flow from land to water.

Offshore and onshore breezes operate day in and day out on sunny days, providing a steady supply of wind energy. Because offshore and onshore winds are fairly reliable, coastal regions of the world are often ideal locations for small (and large) wind turbines.

Coastal winds are more consistent than winds over the interior of continents and also tend to be more powerful because of the relatively smooth and unobstructed surface of open waters. That is to say, wind moves rapidly over water because lakes and coastal waters provide very little resistance to its flow, unlike forests or cities and suburbs, which dramatically lower surface wind speeds.

Mountain-Valley Breezes

Like coastal winds, mountain-valley breezes arise from the differential heating of the Earth's surface. To understand how these winds are formed, let's begin in the morning.

As the Sun rises on clear days, sunrays strike the valley floor and begin heating the ground, valley walls and mountains. As the ground and valley walls begin to warm, the air above them warms. It then expands and begins to flow upward. This process is known as convection. (Convection is the transfer of heat in a fluid or a gas that is caused by the movement of the heated air or fluid itself.) While some of this warm air rises vertically, mountain valleys also tend to channel the solar-heated air through the valley toward the mountains (Figure 2.2). As the warmed air moves up a valley, cooler air from surrounding areas flows in to replace it. This wind is known as a valley breeze.

Throughout the morning and well into the afternoon, breezes flow up-valley — from the valley floor into the mountains. These breezes tend to reach a crescendo in the afternoon. When the Sun sets, however, the winds reverse direction, flowing down valley.

Winds flow in reverse at night because the mountains cool more quickly than the valley floor. Cool, dense air (high-pressure air) from the mountains sinks and flows down through the valleys like

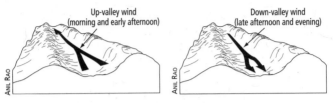

Fig. 2.2a and 2.2b: *Mountain-Valley Breezes. Mountain-valley winds can provide a reliable source of wind power if conditions are just right.* (a) Up-valley winds. (b) Down-valley winds.

the water in a mountain stream, creating steady and often predictable down-valley or mountain breezes.

Together, valley and mountain winds are known as mountain-valley breezes. As a rule, mountain breezes (down-flowing winds) tend to be stronger than daytime valley breezes.

Mountain-valley breezes typically occur in the summer, a time when solar radiation is greatest. They also typically occur on calm days when the prevailing winds (larger regional winds, which will be discussed shortly) are weak or nonexistent.

Mountain-valley winds also form in the presence of prevailing winds — for example, when a storm moves through an area. In such instances, mountain or valley winds may "piggy back" on the prevailing winds, creating even more powerful (and hence higher energy) winds. When consistently flowing in the same direction, such winds can provide a great deal of power that can be tapped to produce an abundance of electricity.

Large-Scale Wind Currents

Local winds can be a valuable source of energy. The winds on which most people rely, however, are those produced by much larger air masses that result from regional and global air circulation. They create dominant wind-flow patterns, known as prevailing winds.

Prevailing winds, like local winds, are created by the differential heating of the Earth's surface, but on a much larger scale. Here's how they are formed: As shown in Figure 2.3, the Earth is divided into three climatic zones: the tropics, temperate zones and poles.

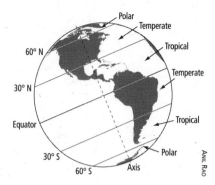

Fig. 2.3: *Climate Zones. The Earth is divided into three climate zones in each hemisphere. In the Northern Hemisphere, warm air from the tropics flows northward by convection, creating the global circulation pattern.*

Because the tropics are more directly aligned with the Sun throughout the year, they receive more sunlight and are, therefore, the warmest regions on Earth.

The temperate zones lie outside the tropics, in both the Northern and Southern Hemispheres. They receive less sunlight than the tropics and so are cooler. The North and South poles receive the least amount of sunlight and are the coolest regions of our planet.

As shown in Figure 2.4a, global air circulation is created by hot air produced in the tropics. This air expands and rises (as in local air circulation patterns). Cool air from the northern regions — as far north as the poles — moves in to fill the void. The result is huge air currents that flow from the poles to the equator.

Although air generally flows from the North and South poles toward the equator, circulation patterns are a bit more complicated. In the Northern Hemisphere, some of the warm air moving northward cools and sinks back to the Earth's surface, as shown in Figure 2.4b. It then flows back toward the equator creating the trade winds. Because the trade winds blow quite consistently, they are a potentially huge and reliable source of energy for residents and nations fortunate enough to lie in the winds' path.

As shown in Figure 2.4b, a substantial amount of the northward-moving air continues on toward the North Pole, traveling over the temperate zone.

The figure also shows air masses flowing northward across the temperate zone split into two, creating higher- and lower-level

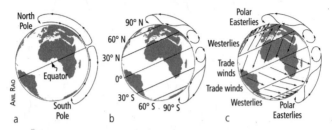

Fig. 2.4: *Global Air Circulation. (a) As shown here, warm tropical air rises and flows toward the poles. Cold polar air flows toward the equator. (b) Warm tropical air loses some of its heat and sinks toward the Earth's surface, then flows back toward the equator, creating the trade winds. Air masses moving over the temperate zone split into upper and lower winds. (c) Wind patterns caused by the Coriolis effect, resulting from the rotation of the Earth on its axis.*

winds. When the upper winds reach the North Pole, this cold air sinks and then flows southward back toward the equator.

If no other forces were at work, winds flowing back to the equator would flow from north to south. As shown in Figure 2.4c, they don't. Other factors influence the movement of air masses across the surface of the planet. One of the most significant is the Earth's rotation, which results in a phenomenon known as the Coriolis effect.

The Coriolis Effect

To understand why prevailing winds deviate from the expected patterns based solely on convection, let's start with the trade winds. As shown in Figure 2.4c, the trade winds in the Northern Hemisphere flow not from north to south, as you might expect, but from the northeast to southwest. Why?

Because they are "deflected" by the Earth's rotation.

In reality, the Earth's rotation doesn't deflect winds. It makes it appear as if the winds have been deflected. The apparent deflection in wind direction in the tropics is a planetary sleight of hand, an illusion produced by the rotation of the Earth on its axis. To understand this phenomenon, consider a simple example. Imagine that you board a plane leaving the North Pole. The pilot plots a course that will take you due south toward an airfield on Sri Lanka just south of India. If the pilot flies the plane due south the entire trip, however, the plane will end up somewhere over the Arabian Sea because of the Earth's rotation.

As shown in Figure 2.5, as the plane travels south, the Earth rotates beneath it. The Earth rotates eastward. If you plot the flight path of the plane it appears to have been deflected. In reality, it only looks that way.

The apparent deflection of the plane's path is the Coriolis effect. In the Northern Hemisphere, the deflection is to the right of the direction of travel. In the Southern Hemisphere, the deflection is to the left.

Winds flowing north or south also appear to be deflected thanks to the Coriolis effect. The south-flowing trade winds, for instance, appear to flow from northeast to southwest.

In the temperate zone, as shown in Figure 2.4c, the low-level north-flowing winds that sweep across the surface of the Earth flow across the North American continent not from south to north but from the southwest to northeast. These are the prevailing south-westerly winds that blow across the Great Plains of North America. Many a wind farm and many a small wind operation depend on them.

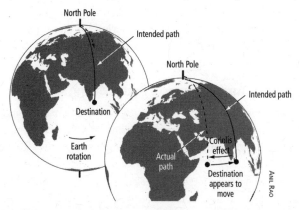

Fig. 2.5: *The Coriolis Effect. The rotation of the Earth causes an apparent deflection in the path of winds. This can be understood by observing the flight of a plane that begins at the North Pole and heads south toward Sri Lanka. The plane appears to veer off course. It hasn't. The Earth's rotation makes it look that way.*

Wind from Storms

Winds are often associated with storms. Storms, in turn, are produced when high-pressure and low-pressure air masses collide. High and low pressure zones move across the continents.

Low-pressure air masses originate in the tropics. They are created by the huge influx of solar energy in these regions. Huge masses of low-pressure air frequently break away and migrate northward, sweeping across the North American continent.

High-pressure air masses originate in the North and South poles, regions of more or less permanent cold, high-pressure air. Like warm tropical air, huge masses of cold Arctic air also break loose and drift southward, sweeping across the Northern Hemisphere.

High-pressure and low-pressure air masses, often measuring 500 to 1,000 miles in diameter, move across continents. The movement of high- and low-pressure air masses across continents is steered by prevailing winds and by the jet stream (high altitude winds). As these air masses collide, they produce an assortment of weather, often accompanied by winds. As with all other forms of wind, storm winds are created by differences in pressure between high- and low-pressure air masses. The greater the difference in pressure between a high-pressure air mass and a "neighboring" low-pressure air mass, the stronger the winds. In some cases, these winds contain an enormous amount of energy.

Friction, Turbulence and Smart Siting

Wind does not flow smoothly over the Earth's surface. It encounters resistance, known as friction. This results in a phenomenon called ground drag. Ground drag is caused by friction when air flows across a surface.

Friction is the force that resists movement of one material against another. You create friction, for example, when you rub your hands together. When wind flows across land or water, friction occurs. This reduces the speed at which air moves over a surface.

Ground drag due to friction, however, varies considerably, depending on the roughness of the surface. The rougher or more irregular the surface, the greater the friction. As a result, air flowing across the surface of a lake generates less friction than air flowing over a meadow. Air flowing over a meadow generates less friction than air flowing over a forest.

Friction extends to a height of about 1,650 feet (500 meters). However, the greatest effects are closest to Earth's surface — the first 60 feet over a relatively flat, smooth surface. Over trees, the greatest effects occur within the first 60 feet (18 meters) above the tree line.

Friction has a dramatic effect on wind speed at different heights. For instance, a 20-mile-per-hour wind measured at 1,000 feet above land covered with grasses flows at 5 miles per hour 10 feet above the surface. It then increases progressively until it breaks loose from the influence of the ground drag or friction. Figure 2.6 shows the difference in wind speed at 50 meters (165 feet) to 5 meters (16.5 feet). Figure 2.7 compares wind speed over a grassy area to wind speed over a forest, a significantly rougher surface. Notice that the wind speed increases more rapidly above the forest.

Ground drag dramatically influences wind speed near the surface of the ground where residential wind generators are located. Because the effects of friction decrease with height above the surface of the Earth, savvy installers typically mount their wind machines on towers 80 to 120 feet high (24 to 37 meters), or even as high as 180 feet (55 meters) in forested regions, so their turbines are out of the most significant ground drag. At these heights, the

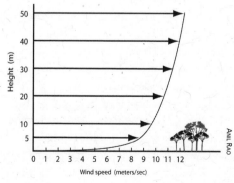

Fig. 2.6: *Effect of Ground Drag. Winds move more slowly at ground level due to friction. Friction diminishes with height, so wind speed increases.*

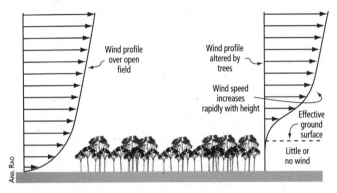

Fig. 2.7: *Wind Speed vs. Height. These graphs compare the wind speed over a grassy area and a forest. As you can see, a forest virtually eliminates ground-level winds. As a result, the effective ground level shifts upward. Note that the wind speed increases more rapidly with height over a forest than over a grassy area. Wind turbines placed well above the tree line can avail themselves of powerful winds.*

winds are substantially stronger than near the ground. As discussed shortly, a small increase in wind speed can result in a substantial increase in the amount of power that's available from the wind and the amount of electricity a wind generator produces. Mounting a wind turbine on a tall tower therefore maximizes the electrical output of the machine. Placing a turbine on a short tower has just the opposite effect. It places the generator in the weaker winds and is a bit like mounting solar panels in the shade.

Another natural phenomenon that affects the output of most wind turbines is turbulence. Turbulence is produced as air flowing across the Earth's surface encounters objects, such as trees or buildings. They interrupt the wind's smooth laminar flow, causing it to tumble and swirl, the same way rocks in a stream interrupt the flow of water (Figure 2.8). Rapid changes in wind speed occur behind large obstacles and winds may even flow in the direction opposite to the wind. This highly disorganized wind flow is referred to as turbulence.

Turbulent wind flows wreck havoc on wind machines, especially the less expensive, lighter-weight wind turbines often installed on

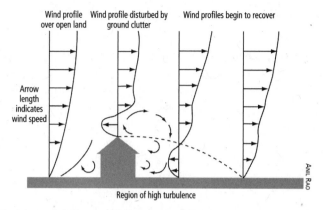

Wind profile over open land Wind profile disturbed by ground clutter Wind profiles begin to recover

Arrow length indicates wind speed

Region of high turbulence

Fig. 2.8: *Ground Clutter and Turbulence. Trees, houses, barns, silos, billboards, garages and other structures are referred to as ground clutter. They create turbulence. Like eddies behind rocks in streams, the turbulent zone contains a fluid (air) that swirls and tumbles, moving in many directions. Turbulence reduces the harnessable energy of the wind and causes more wear and tear on wind machines, damaging them over time.*

short towers by cost-conscious homeowners. Buffeted by turbulent winds, wind machines hunt around on the top of their towers, constantly seeking the strongest wind, starting and stopping repeatedly. This decreases the amount of electricity a turbine generates.

Turbulence also causes vibration and unequal forces on the wind turbine, especially the blades, that may weaken and damage the machine. Turbulence, therefore, increases wear and tear on wind generators and, over time, can destroy a turbine. The cheaper the turbine, the more likely it will be destroyed in a turbulent location. A homeowner may find that a machine he or she had hoped would produce electricity for ten to twenty years only lasted two to four years.

Turbulence is to a wind machine like potholes to your car.
— Robert Preus, Abundant Renewable Energy

When considering a location to mount a wind turbine, be sure to consider turbulence-generating obstacles such as silos, trees, barns, houses and other wind turbines. Proper location is the key to avoiding the damaging effects of turbulence. Turbulence can also be minimized by mounting a wind turbine on a tall tower. In sum, then, mounting a wind generator on a tall tower offers four benefits: (1) it situates the wind generator in the stronger higher-energy-yielding winds, substantially increasing electrical production, (2) it raises the machine out of damaging turbulent winds, (3) it decreases the wind turbine's maintenance and repair requirements, and (4) it increases the wind turbine's useful lifespan substantially, perhaps tenfold. Longer turbine life means less overall expense — and more electricity from your investment.

As shown in Figure 2.9, all obstacles create a downstream zone of turbulent air, or "turbulence bubble." It typically extends vertically about twice the height of the obstruction and extends downwind approximately 15 to 20 times the height of the obstruction. A 20-foot-high house creates a turbulence bubble that extends 40 feet above the ground and 300 to 400 feet downwind. As illustrated in Figure 2.9, the turbulence bubble also extends upwind — about two times the object's height. In this case, the upwind bubble extends about 40 feet upwind from the house. The upstream portion of the bubble is created by wind backing up as it strikes the obstacle — much like water flowing against a rock in a river.

To avoid costly mistakes, installers recommend that wind machines be mounted so that the complete rotor (the hub and the blades) of the wind generator is *at least* 30 feet (9 meters) above the closest obstacle within 500 feet (about 150 meters), or a tree line in

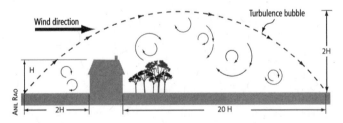

Fig. 2.9: *Turbulence Bubble. The turbulence bubble extends upwind, downwind, and even above the clutter. The bubble shifts when wind direction changes.*

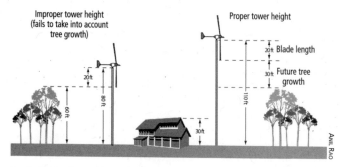

Improper tower height
(fails to take into account
tree growth)

Proper tower height

20ft Blade length

30ft Future tree
growth

20ft

60 ft

80 ft

110 ft

30ft

ANIL RAO

Fig. 2.10: *Treelines and Wind Speed. Siting a wind turbine in an open area surrounded by trees is possible. As explained in the text, the top of the tree line is the effective ground level. The turbine needs to be mounted well above the trees to perform optimally.*

the area, whichever is higher (Figure 2.10). Don't listen to those who recommend lesser heights. Many unhappy customers will attest to that!

If your home or business is in an open field surrounded by trees, the wind turbine needs to be well above the tree line (Figure 2.10). Remember, too, to account for growth of trees over the 20- to 30-year life span of your wind system when determining tower height.

The Mathematics of Wind Power

To understand how important it is to mount a wind turbine on a tall tower, consider a simple mathematical equation. It's called the power equation and is used to calculate the power available from the wind. This equation shows us that three factors influence the output of a wind energy system: (1) air density, (2) swept area, and (3) wind speed — all explained shortly.

The power equation is: $P = \frac{1}{2} d \times A \times V^3$.

P stands for the power available in the wind (not the power a wind generator will extract — that's influenced by efficiency and other factors). Density of the air is d. Swept area is A. Wind speed is V.

Air Density

Air density is the weight of air per unit volume, which varies with elevation. As a general rule, anticipate a decrease in the air density of about 3 percent per 1,000 feet (300 meters) increase in elevation. As a result, air density doesn't affect the power available from the wind until elevation reaches 2,500 feet (760 meters) above sea level. At 3,000 feet (about 910 meters) above sea level, the air density is 9 percent lower than at sea level. At 5,000 feet (about 1,525 meters) air density declines by about 15 percent.

Air density is also a function of relative humidity, although the difference between a dry and humid area is usually negligible.

Temperature also affects the density of air. Warmer air is less dense than colder air. Consequently, a wind turbine operating in cold (denser) winter winds would produce slightly more electricity than the same wind turbine in warmer winds blowing at the same speed.

Although temperature and humidity affect air density, they are not factors we can change. Installers must be aware of the reduced energy available at higher altitudes, however, so they don't create unrealistic expectations for a wind system.

Although density is not a factor we can control, wind installers do have control over a couple of other key factors, notably, swept area (A) and wind speed — both of which have a much greater impact on the amount of power available to a wind turbine and the electrical output of the machine than air density.

Swept Area

Swept area is the area of the circle that the blades of a wind machine create when spinning. It is a wind machine's collector surface. The larger the swept area, the more energy a wind turbine can capture from the wind. Swept area is determined by blade length. The longer the blades, the greater the swept area. The greater the swept area, the greater the electrical output of a turbine. As the equation suggests, the relationship between swept area and power output is linear. Theoretically, a ten percent increase in swept area will result in a ten percent increase in electrical production. Doubling the swept area doubles the output.

When shopping for a wind turbine, always convert blade length to swept area, if the manufacturer has not done so for you (they usually do). Swept area can be calculated using the equation A = $\pi \cdot r^2$.

In this equation, A is the area of the circle, the swept area of the wind turbine. The Greek symbol is pi, which is a constant: 3.14. The letter r stands for the radius of a circle, the distance from the center of the circle to its outer edge. For a wind turbine, radius is usually about the same as the length of the blade.

Because swept area is a function of the radius squared, a small increase in radius, or blade length, results in a large increase in swept area. As an example, a wind generator with an 8-foot blade has a swept area of 200 square feet. A wind generator with a 25 percent longer blade, that is, a 10-foot blade, has a 314 square-foot swept area. Thus, a 25 percent increase in blade length results in a 57 percent increase in swept area and, theoretically, a 57 percent increase in electrical production.

Wind Speed

Although swept area is more important than density, wind speed is even more important in determining the output of a wind turbine. That's because the power available from the wind increases with the cube of wind speed. This relationship is expressed in the power equation as V^3 or V x V x V — wind speed multiplied by itself three times.

Consider an example: Suppose that you mount a wind machine 18 feet (5.5 meters) above the ground surface on the grasslands of

Shopping Tip

Because swept area is such an important determinant of the output of a wind turbine, we strongly recommend focusing more on the swept area of a wind turbine than on its rated power — at least until the industry can come up with a standardized way of measuring and reporting rated power.

Nebraska. Suppose that the wind is blowing at eight miles per hour (3.6 meters per second). A friend, who knows how important it is to mount a wind machine on a tall tower, installs an identical wind turbine on a 90-foot (27-meter) tower. When the wind is blowing at 8 mph where your turbine flies at 18 feet, an anemometer on your friend's 90-foot tower indicates that the wind is blowing at 10 miles per hour. Wind speed is 25 percent higher. What's the difference in available power?

The power available in the wind can be approximated by multiplying the wind speeds by themselves three times. (Units aren't important for this comparison.) For the lower turbine the result is 8 x 8 x 8 or 512. The power available to the wind turbine mounted on a 90-foot tower is 10 cubed or 10 x 10 x 10 which is 1,000. Thus, a two-mile-per-hour increase in wind speed, a paltry 25 percent increase, doubles the available power. Put another way, a 25 percent increase in wind speed yields an increase of nearly 100 percent. The important lesson is that because power is function of V^3, a small increase in wind speed results in a very large increase in the power available to a wind turbine. This can result in a very large increase in the electrical output of a wind turbine.

Although winds are out of our control, homeowners can affect the wind speed at their wind turbines by choosing the best possible site and by installing their machines on the tallest towers.

Closing Thoughts

Wind is a tremendous resource available in many parts of the world thanks to the Sun's unequal heating of the Earth's surface. Depending on your location, you may be able to take advantage of offshore and onshore winds or perhaps mountain-valley winds. Prevailing winds and winds that are created between high-pressure and low-pressure air masses could become your ally in reducing your carbon footprint and meeting your own needs sustainably. Don't forget to take into account ground drag that slows wind speed and robs you of additional energy and damages your wind turbine. Site wisely and you'll be repaid day after day after day.

CHAPTER 3

WIND ENERGY SYSTEMS

Wind-electric systems fit into three categories: (1) grid-connected, (2) grid-connected with battery backup, and (3) off-grid. In this chapter, we'll examine each system and discuss the pros and cons of each. We'll also examine hybrid systems, consisting of a wind turbine plus another form of renewable energy. This information will help you decide which system suits your needs and lifestyle. To begin, let's take a look at two of the main components of wind systems, wind turbines and towers. Subsequent chapters contain more detailed discussions of these and other components.

Wind Turbines

Most wind turbines in use today are horizontal axis units, or HAWTs, (explained shortly) with three blades attached to a central hub. Together, the blades and the hub form the rotor. In many wind turbines, the rotor is connected to a shaft that runs horizontal to the ground, hence the name. It is connected to an electrical generator. When the winds blow, the rotor turns and the generator produces alternating current (AC) electricity. (See the accompanying box for an explanation of AC electricity.)

One of the key components of a successful wind generator is the blades. They capture the wind's kinetic energy and convert it into mechanical energy (rotation). It is then converted into electrical energy by the generator.

The generators of wind turbines are often protected from the elements by a durable housing made from fiberglass or aluminum (Figure 3.1a). However, in many modern small wind turbines, the generators are exposed to the elements (Figure 3.1b).

Most wind turbines in use today have tails that keep them pointed into the wind to ensure maximum production. However,

AC vs. DC Electricity

Electricity comes in two basic forms: direct current and alternating current. Direct current (DC) electricity consists of electrons that flow in one direction through the electrical circuit. DC electricity is the kind produced by flashlight batteries or the batteries in cell phones, laptop computers, or portable devices such as iPods.

Most wind turbines produce alternating current electricity. In alternating current, the electrons flow back and forth. That is, they switch, or alternate, direction in very rapid succession, hence the name. Each change in the direction of flow (from left to right and back again) is called a cycle.

In North America, electric utilities produce electricity that cycles back and forth 60 times per second. It's referred to as 60-cycle-per-second — or 60 hertz (Hz) — AC. The hertz unit commemorates Heinrich Hertz, the German physicist whose research on electromagnetic radiation served as a foundation for radio, television and wireless transmission. In Europe and Asia, the utilities produce 50-cycle-per-second AC.

AC electricity is also produced by electrical generators in hydroelectric and power plants that run on fossil fuels or nuclear fuels. No matter what form of energy is used to turn a generator, all of them operate on the principle of magnetic induction — they move coils of copper wire through a magnetic field (or vice versa). This causes electrons to flow through the coils, producing electricity.

a   b

Fig. 3.1a and 3.1b: *Wind Turbine Design. (a) The generators in many small wind turbines are housed in a protective case made from aluminum or fiberglass. (b) Others, like this one, are not.*

some very successful turbines like those made by the Scottish company Proven (pronounced PRO-vin) are designed to orient themselves to the wind without tails. (More on them in Chapter 5.)

Towers

Another vital component of all wind systems is the tower, discussed in more detail in Chapter 6. Residential wind generator towers come in three varieties: (1) freestanding, (2) fixed guyed, and (3) tilt-up (Figure 3.2).

Freestanding towers may be either monopoles or lattice structures. Freestanding monopole towers consist of high-strength hollow tubular steel like that used for streetlight poles. Lattice towers consist of tubular steel pipe or flat-metal steel bolted or welded together to form a lattice structure like the Eiffel Tower or transmission towers used for high-voltage transmission lines.

Freestanding towers must be strong enough to support the weight of the wind turbine, but more importantly, they need to be strong enough to withstand the forces of the wind acting on the turbine and the tower itself. Towers must have a large foundation

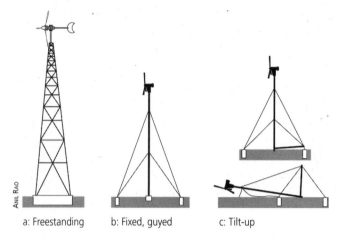

a: Freestanding   b: Fixed, guyed   c: Tilt-up

Fig. 3.2: *Tower Types. Three types of towers in use today: (a) freestanding, (b) fixed guyed, and (c) tilt-up.*

to counteract the tremendous forces applied by the wind, forces that could easily topple a weak or poorly anchored tower. Large amounts of steel and concrete are required to accomplish this task. This makes freestanding towers the most expensive tower option.

The second tower type is the fixed guyed tower (Figure 3.2b). Guyed towers may be made of high-strength steel pipe or may be lattice structures supported by high-strength steel cables, known as guy cables or guy wires. Guy cables extend from anchors in the ground to tower attachment points. They stiffen towers so they don't buckle and hold them in place so they don't lean or fall over.

Because guy cables strengthen and stabilize the tower, less steel is required for them than in freestanding towers. Guy cables are attached to the tower every 20 to 30 feet, although wider spacing is also possible. A 120-foot-tall tower may require four sets of cables.

The third type of tower is the tilt-up tower, so named because it can be raised and lowered — tilted up and down. This makes it possible to lower a turbine so it can be inspected, maintained and repaired at ground level. Tilt-up towers typically consist of high-strength steel pipe or a lattice structure. Both are supported by guy cables.

Towers are as important as the wind turbine itself. Unfortunately, many wind turbines are mounted on towers that are much too short. This occurs out of ignorance and misguided frugality. Installing a turbine on a short tower is a common and costly mistake. Properly sized towers place wind turbines in the path of more powerful winds and raise turbines above turbulence created by ground clutter, which severely diminishes the quality and quantity of wind. Don't let anyone talk you into a short tower!

Electricity produced by a wind turbine runs down wires attached to the tower. The wires typically run externally — for example, alongside a tower leg of a lattice tower — to a junction box at the base of the tower. From there, they typically run underground in conduit to the point of use.

As you shall soon see, wind-electric systems involve a number of additional components. We'll describe them as we explore the three main types of wind energy system.

Wind Energy System Options

As noted in the introduction, wind systems fall into three main categories: (1) grid-connected, (2) grid-connected with batteries, and (3) off-grid. We'll begin with the simplest, the grid-connected system.

Grid-Connected Systems

Grid-connected systems are so named because they connect directly to the electrical grid. They are also referred to as batteryless grid-connected or batteryless utility-tied systems because they do not employ batteries to store surplus electricity[1].

In batteryless grid-tie systems, the electrical grid accepts surplus electricity — electricity produced by the turbine in excess of

---

1. Because the term "grid" refers to the high-voltage transmission system, not local electrical distribution systems, customers are connected to the grid through their local utility and indirectly to its distribution system, the grid. Because of this, the terms "grid-connected" and "grid-tied," which are used by the renewable energy community, are not entirely accurate. "Utility-connected" or "utility-tied" would be better terms, but that is not the common usage in the renewable energy world.

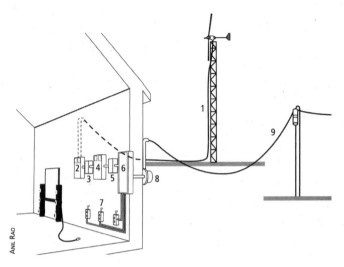

1. Electric wire from turbine carries wild AC electricity to house
2. DC control panel
3. DC disconnect
4. Inverter
5. AC disconnect
6. Breaker box
7. Electrical outlets
8. Electrical meter
9. Electric wire carries AC electricity to and from house

Fig. 3.3: *Grid-Connected Wind System. The grid-connected wind system is the simplest of all systems. Wild AC electricity produced by the turbine is first fed into the controller. The inverter produces grid-compatible AC electricity to power household loads. Surpluses are backfed onto the electrical grid.*

demand. When a wind system is inactive, the grid supplies electricity to the home or business. The grid therefore serves as the storage medium.

As shown in Figure 3.3, a batteryless grid-connected system consists of six main components: (a) a wind generator specifically designed for grid connection, (b) a tower, (c) an inverter/power conditioner, (d) a main service panel (e) meters, and (f) safety disconnects.

In most batteryless grid-connected systems, the wind generator produces "wild AC" electricity. Wild AC is alternating current electricity the frequency and voltage of which vary with wind speed.

Frequency is the number of times electrons switch direction every second and, as noted earlier, is measured in hertz, or cycles per second.

The flow of electrons through an electrical wire is created by an electromotive force that scientists call voltage. Voltage is electrical pressure, the driving force that causes electrons to move. The unit of measurement for voltage is volts. In most small wind turbines, the faster the blades spin, the higher the voltage and the greater the frequency.

Wild AC, produced by wind turbines, is not directly usable. Appliances and electronic devices require a tamer version of electricity — alternating current with a fairly constant frequency and voltage, like that available from the grid. In a grid-connected system, the wild voltage must first be "tamed." That is, its frequency and voltage must be converted to standard values. This occurs in two additional components, the controller and inverter (Figure 3.3). The inverter converts the electricity to grid-compatible AC — 60 cycle per second, 120-volt (or 240-volt) electricity. Because the inverter produces electricity in sync with the grid, it's often referred to as a synchronous inverter.

While grid-compatible wind generators typically produce wild AC, another type of wind generator is also found in the small wind market. It is an induction generator. As explained more fully in Chapter 5, an induction generator produces grid-compatible AC electricity without a controller or inverter.

The 120-volt or 240-volt AC produced by the inverter (or directly by an induction generator) flows to the breaker box, which is where the circuit breakers are found. From here, the electricity flows along wires in a house or business to electrical devices drawing power. If the wind machine is producing more electricity than is needed, the excess is fed onto the grid.

Surplus electricity backfed onto the grid travels from the main service panel through the utility's electric meter, typically mounted on the outside of the building. It then flows through the wires that connect to the grid. The surplus electricity then travels along the power lines where it flows into neighboring homes or businesses.

A utility electric meter monitors the electricity fed onto the grid so the utility can credit the producer for its contribution. The meter also keeps track of electricity the power company supplies to homes and businesses when their wind systems are not generating. To learn how the electric company measures what you are putting onto the grid and how they "pay" for it, check out the accompanying box, "Net Metering in Grid-Connected Systems."

In addition to the electric meter — or meters (some utilities require two or more meters) — that monitor the flow of electricity onto and off the local utility grid, grid-connected wind energy systems often contain safety disconnects. These are manually operated switches that enable service personnel to disconnect at a couple of key points in the system to prevent electrical shock if service is required. As shown in Figure 3.3, an AC disconnect is located between the inverter and breaker box. When switched off, it disconnects and isolates the wind energy system from the household circuits and the grid. The AC disconnect must be mounted outside so that it is accessible to utility company personnel so they can isolate the wind system from the grid when working on the electric lines in your area without fear of shock — for instance, if a line goes down in an ice storm. The AC disconnect must also contain a switch that can be locked in the off position by the utility worker so that the homeowner or a family member doesn't accidentally turn the system back on prior to the completion of repairs.

Lockable AC disconnects are required by most utilities. However, many experienced electric companies like the large California utilities and Colorado's Exel, which collectively have thousands of solar- and wind-electric systems connected to their systems, have dropped the requirement for utility company-accessible, lockable disconnects. The California utilities have found that they can't use them because there are too many to disconnect at once. More importantly, these companies have come to realize that they're simply not needed. That's because grid-compatible inverters automatically shut off when the utility power goes down. As a result, no electricity can flow onto the grid. A properly installed

Net Metering in Grid-Connected Systems

Utilities keep track of the two-way flow of electricity from grid-connected renewable energy systems in one of two ways. In many cases, they install a single dial-type electric meter. This meter measures kilowatt-hours of energy and can run forward or backward. It runs forward (counting up) when your home or business uses more energy than your wind turbine is producing. It runs backward (counting down) when the wind turbine produces more electricity than is being used.

In many new installations, utilities install digital meters that tally electricity delivered to and supplied by a home or business. They don't "run backward," as older, dial-type meters do. They simply keep track of electricity coming from and going to the grid so the utility can determine whether you've produced as much, more, or less than you've consumed.

Still other utilities install two meters, one to tally the electricity delivered to the grid and another to keep track of electricity supplied by the grid. (They typically charge to install the second meter and may charge a separate monthly fee to read it.)

Most companies employ a billing system known as net metering. Net metering is a system in which the electric bill is based on net consumption — consumption minus production. That is to say, a customer's electric bill is based on the amount of utility energy consumed minus the amount of energy provided to the grid from a renewable energy system. The "net" in net metering refers to net kilowatt-hours.

In a net metering arrangement with two meters, one meter tracks consumption while the other tallies a customer's contribution to the grid. The customer is charged for the electricity he or she consumes and is credited for the electricity he or she feeds onto the grid at the same ☛

rate (retail rate) up to the point where there's a surplus backfed onto the grid. The differences in net metering programs in different states stem from differences in ways utilities treat surpluses (net excess generation or NEG).

Net excess generation can be reconciled at the end of the month (monthly net metering) or at the end of the year (annual net metering).

If you're thinking that wind could be a profitable venture, don't get your hopes up. While a few utilities pay retail rates, most reimburse for net excess generation at their *wholesale* rate — that is, what they pay to generate electricity. Many utilities don't write checks at all. They simply "take" the surplus without payment to the customer, like cell phone companies do. It all depends on state law.

As a rule, monthly net metering is generally the least desirable option, especially if surpluses are "donated" to the utility company or reimbursed at wholesale rates (a.k.a. avoided cost). Annual reconciliation is a much better deal. It permits wintertime surpluses to be "banked" to offset summertime shortfalls, if any.

The ideal arrangement from a customer's standpoint is to negotiate a continual month-to-month roll-over, so there's no concern about losing credit for your wind-powered kilowatt-hours. We're only aware of one state in which credits carry forward indefinitely, and that's Kentucky.

Net metering is mandatory in many states. At this writing (May, 2009), 41 states and the District of Columbia have instituted net metering, although there are substantial differences among them. The programs vary with respect to which utilities are required to participate in the program, the size and types of systems that qualify, payment or lack of payment for net excess generation (wholesale, retail or no payment), and so on.

Some states only require "investor-owned" utilities to offer net metering. These companies typically serve ☛

urban regions where wind energy systems are not well suited. In many states, municipal power companies and rural cooperatives are exempt from net metering laws. The exemption of rural co-ops from net metering is unfortunate, because rural areas are more likely to have the best wind resources, less ground clutter, fewer zoning restrictions, and homes on larger acreage, which are ideal for installing wind turbines.

Utilities that don't offer net metering may use another system called *buy-sell* or *net billing*. In these arrangements, utilities typically install two meters, one to track electricity the utility sells to the customer, and another to track electricity the customer feeds onto the grid. At first blush, this arrangement may seem very similar to net metering, however, it is quite different. Unlike net metering, utilities that engage in buy-sell arrangements typically charge their customers retail rates for all the electricity they draw from the grid but pay customers wholesale rates for *all* the electricity fed onto the grid by renewable energy systems. For example, a utility may charge 10 to 15 cents per kilowatt-hour for electricity they supply to the customers, but pay wholesale rates of 2 to 3 cents per kilowatt-hour for *all* the electricity that customers deliver to the grid. How does this work out financially for the small-scale producer?

As you might suspect, not very well.

Suppose that a customer consumed 500 kilowatt-hours of electricity from the grid but fed 1,000 kilowatt-hours of electricity onto the grid in December, a typically windy month. Let's suppose that the retail rate for electricity was 10 cents per kilowatt-hour and the generation and delivery costs (that is, the wholesale cost) came to 3 cents per kilowatt-hour. In a buy-sell agreement, the utility would charge the customer 10 cents per kilowatt-hour for all the electricity delivered to them, or $50 for the 500 kilowatt-hours. The utility would then credit the customer 3 cents per kilowatt-hour for the 1,000 kilowatt-hours of ☞

electricity delivered to the grid. That is, the customer would be paid or credited $30.00 for the 1,000 kilowatt-hours of electricity sold to the utility. As a result, the customer would end up owing the utility $20 plus a meter-reading fee of $10 to $20.

Although (at this writing) nine states do not have net metering laws for wind, progress in this area is moving quickly. More and more utilities have come to realize that it is often cheaper — and less hassle — for them to net meter than to install two meters. For information about net metering rules for individual states, go to the Database for State Incentives on Renewables and Efficiency (dsireusa.org). Click on your state and scroll down to net metering. ∎

grid-connected wind-electric system will not backfeed a dead grid. Period.

The Pros and Cons of Grid-Connected Systems

Batteryless grid-connected systems represent the majority of all new wind systems in the United States. Their pluses and minuses are summarized in Table 3.1.

On the positive side, batteryless grid-connected systems are relatively simple and typically the least expensive option — often 25% cheaper than battery-based systems. They also require less maintenance than battery-based systems.

Another substantial advantage of these systems is that they can store an unlimited amount of electricity on the grid (so long as the grid is operational). Although grid-connected systems don't physically store excess electricity on site like a battery-based system for later use, they "store" surplus electricity on the grid in the form of a credit on your utility bill. When winds fail to blow — or a wind turbine isn't producing enough electricity to meet demand — electricity is drawn from the grid, using up the credit. Unlike a battery bank, you can never "fill up" the grid. It will accept as much electricity as you can feed it.

By crediting a producer for electricity fed onto the grid, a utility says, "You've supplied us with *x* kilowatt-hours of electricity. When you need electricity, we'll supply you with an equal amount at no cost. If at the end of the month you've supplied more than you consume, we'll either pay you for it or carry the surplus over to the next month."

Another advantage of grid storage is utility storage of electricity is not subject to losses that occur when electricity is stored in a battery. As described in Chapter 7, when electricity is stored in a battery, it is converted to chemical energy. When electricity is needed, the chemical energy is converted back to electrical energy. As much as 20 to 30 percent of the electrical energy fed into a battery bank is lost due to conversion inefficiencies and other factors. In sharp contrast, electricity stored on the grid comes back in full. If you deliver 100 kilowatt-hours of electricity, you can draw off 100 kilowatt-hours. (The grid has losses too, however, net metered customers get 100 percent return on their stored electricity.)

Another advantage of grid-tie systems is that they are greener than battery-based systems. Although utilities aren't the greenest entities in the world, they are arguably greener than battery-based systems. Battery production requires an enormous amount of energy and raw materials. Batteries also contain highly toxic sulfuric acid. Although old lead-acid batteries are recycled, they're often recycled under abysmally poor conditions in less developed countries, exposing employees (often young children) and the environment to toxic chemicals.

Grid-tie systems, when net metered, can provide some income. In windy sites, they may produce surpluses month after month. If the local utility pays for surpluses at retail rates, the surpluses can generate income that helps reduce the cost of the system and the annual cost of producing electricity.

On the downside, grid-connected systems may require extensive negotiations with local utilities. This, though, may become a thing

Table 3.1 Pros and Cons of Batteryless Grid-Tie Systems	
Pros	**Cons**
Simpler than other systems	Vulnerable to grid failure unless an uninterruptible power supply and/or generator is installed
Less expensive	
Less maintenance	
More efficient than battery-based systems	
Unlimited storage of surplus electricity	
Greener than battery-based systems	

of the past. Although some utilities may throw up roadblocks, more and more are becoming cooperative as they become more comfortable with these systems.

Another downside of these systems is that when the grid goes down, so does a batteryless grid-connected wind system. Even when winds are blowing, batteryless grid-tied wind energy systems shut down if an electric line comes crashing down in an ice storm or lightning strikes a nearby transformer, both of which result in a power outage. Even though the winds are blowing, you'll get no power from your system.

If power outages are a recurring problem and outages occur for long periods, you may want to consider installing a standby gas or diesel generator that switches on automatically when the grid goes down. Because a backup generator takes many seconds to start up and come on line, you may want to consider installing an uninterruptible power supply (UPS) on critical equipment such as computers. A UPS contains a battery and a small inverter. If the utility power goes out, it supplies power instantly until its battery runs low. Another alternative is to install a grid-connected system with battery backup, discussed next. In these systems batteries provide backup power to a home or business when the grid goes down.

Grid-Connected Systems with Battery Backup

Grid-connected systems with battery backup are also known as battery-based utility-tied systems. These systems ensure a continuous supply of electricity, even when freezing rain wipes out the electrical supply to your home or business. Figure 3.4 shows the components of these systems: (1) a wind turbine on a tower, (2) a charge controller, (3) an inverter, (4) safety disconnects, (5) breaker box or main service panel, and (6) meters to keep track of electricity delivered to and drawn from the grid.

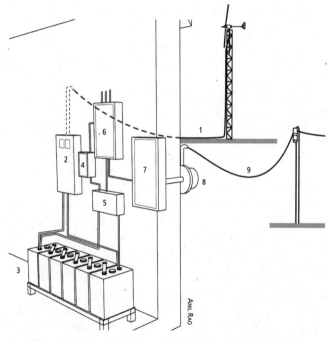

1. Carries wild AC from turbine
2. Controller
3. Battery bank
4. Transfer switch
5. Inverter
6. Subpanel (for critical loads)
7. Main breaker box (all household loads)
8. Utility meter
9. Service line

Fig. 3.4 : *Grid-Connected Wind System with Battery Backup.*

Although grid-connected systems with battery backup are similar to batteryless grid-connected systems, they differ in several ways. The most obvious difference is that battery-based grid-connected systems contain a bank of batteries. They also require a different type of inverter. These systems also contain a meter that monitors the flow of electricity into and out of the battery bank and a device known as a *charge controller*.

Batteries for grid-connected systems with battery backup are either flooded lead-acid batteries or sealed lead-acid batteries. Battery banks in grid-connected systems are typically smaller than those in off-grid systems because they are usually sized to provide sufficient storage to run a handful of critical loads for a day or two until the utility company restores electrical service. Critical loads might include a refrigerator and freezer, a few lights, a well pump, and the blower of a furnace or the boiler and pump in a radiant heating system.

Keeping batteries fully charged is a high priority in these systems. Battery banks are maintained at full charge day in and day out to ensure a ready supply of electricity should the grid go down. It's only when the batteries are topped off and a household's demands are being met that excess electricity is backfed onto the grid.

Batteries are called into duty only when the grid goes down. They're a backup power source. They're not there to supply additional power to run loads that exceed the wind system's output. When demand exceeds supply, electricity is supplied by the electrical grid, not the batteries. When the winds are dead, the grid, not the battery bank, becomes the power source.

Maintaining a fully charged battery bank requires a fair amount of electricity. That's because batteries self-discharge when sitting idly by. Thus, a good portion of the surplus electricity a wind system generates may be devoted to keeping batteries full. Keeping batteries topped off consumes 5 to 10 percent of a system's daily output. (In systems with a low-efficiency and technologically unsophisticated inverter and a large or older battery bank, consumption can be as high as 25 to 50 percent.)

Battery banks in grid-connected systems don't require careful monitoring like those in off-grid systems, but it is a very good idea to keep a close eye on them — just to be sure they'll be functional when the grid goes down. Owners can monitor batteries through a meter that indicates the total amount of electricity stored in the battery bank at any one time. These meters give readings in amp-hours or kilowatt-hours. What do these terms mean?

As most readers know, electricity is the flow of electrons through a wire. Like water flowing through a hose, electricity flows through conductors at varying rates. The rate of flow depends on the voltage.

The flow of electrons through a conductor is measured in amperes or amps for short. An amp is 6.24×10^{18} electrons passing by a point on a conductor per second. The greater the amperage, the faster the electrons are flowing.

One amp of electricity flowing through a wire for an hour is one amp-hour. This term is also frequently used to define a battery's storage capacity. A flooded lead-acid battery, for example, might store 420 amp-hours of electricity. Amp-hours can also be converted to kilowatt-hours, as explained shortly.

Meters also typically display battery voltage. Battery voltage can provide a very general approximation of the amount of energy in a battery — if you know how to interpret this parameter. We'll discuss this topic in Chapter 7.

Another component found in wind energy systems with battery backup is the charge controller, shown in Figure 3.4. Charge controllers contain a component known as a rectifier. It converts AC from the wind generator to DC electricity. It is then fed into the batteries.

Charge controllers also monitor battery voltage. They use this information to protect batteries from being overcharged — having too much electricity driven into them. Overcharging can permanently damage the lead plates in batteries, dramatically reducing battery life.

When the charge controller sees that the batteries are fully charged, it terminates the flow of electricity to them. Surplus

Fig. 3.5: *Dump Load. Resistive heaters like this one are used as dumps for surplus electricity from off-grid wind energy systems.*

electricity is then fed onto the grid, or if the grid is not operational, to a diversion or dump load. Diversion loads are typically resistance-type devices that convert surplus electricity into heat. They are installed in water heaters or as separate space heaters in the basement or a nearby utility room and help put to use the surplus electricity (Figure 3.5).

Pros and Cons of Grid-Connected Systems with Battery Backup

Grid-connected systems with backup power protect against utility failures, although typically only a handful of critical loads can be run from a battery bank. These systems allow homeowners to heat their homes, keep food cold, power a radio, and run emergency medical equipment. In businesses, they protect computers and other vital equipment required to continue operations.

Grid-connected systems with battery backup do have some drawbacks. They cost more to install and operate than batteryless

| Table 3.2 Pros and Cons of a Battery-Based Grid-Tie System ||
Pros	Cons
Provides a reliable source of electricity	More costly than batteryless grid-connected systems
	Less efficient than batteryless grid-connected systems
	Less environmentally friendly than batteryless systems
	Requires more maintenance than batteryless grid-connected systems

grid-connect systems. Flooded lead-acid batteries used in these systems require periodic maintenance and replacement every five to ten years, whether they're used or not. Keeping batteries topped off can also consume a fair amount of a system's daily electrical output.

When contemplating a battery-based grid-tie system, ask yourself three questions: (1) How frequently does the grid fail in your utility's service area? (2) What are your critical loads and how important is it to keep them running? (3) How do you react when the grid fails?

If the local grid is extremely reliable, you don't have medical support equipment to run or need computers for critical financial transactions, and you don't mind using candles on the rare occasions when the grid goes down, why buy, maintain, and replace costly batteries? See Table 3.2 for a quick summary of the pros and cons of battery-based grid-connected systems.

Off-Grid (Stand-Alone) Systems

Those who want to or must supply all of their needs through wind energy or a combination of wind and solar and don't want to be connected to the grid install off-grid systems. As shown in Figure 3.6, this system bears a remarkable resemblance to a grid-connected system with battery backup.

The main source of electricity in an off-grid system is a battery-charging wind turbine. These turbines produce wild AC electricity that is converted (rectified) to DC electricity by rectifiers located in the charge controller.

The controller delivers DC electricity to the battery bank. When electricity is needed, it is drawn from the battery bank via the inverter. The inverter converts the DC electricity from the battery bank, typically 24 or 48 volts, to higher-voltage AC, either 120 or 240 volts, required by households and businesses. The AC then flows to active circuits in the house via the breaker box.

Although off-grid systems resemble grid-connected systems with battery banks, there are some noticeable differences. The first and most obvious is that there are no power lines running from the house or business to the grid. In these systems, then, the wind turbine produces all of the electricity required to meet the owner's needs. Surplus generated during windy periods is stored in batteries for use during low- or no-wind periods. If the batteries are full, the surplus is typically sent to the diversion load.

Off-grid systems are also typically equipped with another source of electricity, often a PV array or a gasoline or diesel generator (gen-set). They help make up for shortfalls.

Off-grid systems also require safety disconnects to permit servicing. A DC disconnect is located between the charge controller and inverter. These systems also contain charge controllers to protect the batteries from overcharging and a low-voltage disconnect to prevent deep discharge of the battery bank.

Off-grid wind energy systems are the most complex of all options. Some systems contain DC circuits. These circuits are fed directly from the battery bank, bypassing the inverter, to power DC lights or refrigerators. Bypassing the inverter saves energy, because inverters are not 100 percent efficient. It takes a little energy to convert DC to AC — usually about 5 to 10 percent.

DC appliances are generally small, difficult to find, expensive, and not always that reliable or as fully equipped as AC appliances. DC refrigerators, for example, do not come with the features that

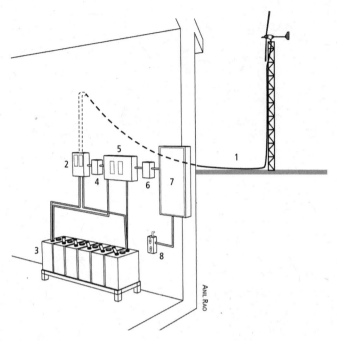

1. Electric wire carries wild AC to controller
2. Controller
3. Battery bank
4. DC disconnect
5. Inverter
6. AC disconnect
7. Main breaker box
8. AC circuits

Fig. 3.6: Most wind turbines in off-grid wind systems produce AC electricity that's converted to DC electricity by the controller. The inverter draws electricity from the batteries, converting it into AC electricity for household use.

many individuals expect, such as automatic defrost or ice makers. DC circuits also require larger, more costly wires and special receptacles. Moreover, the energy lost as low-voltage DC electricity flows through wires is about the same as the losses in an inverter.

To simplify installation of battery-based systems, you may want to consider installing a power center (Figure 3.7). Power centers contain many of the essential components of a renewable energy system, including one or more inverters, the meters needed to monitor

OUTBACK

Fig. 3.7: *Power Center. Power centers like this one contain all of the components needed for a successful installation, all mounted on one panel. They're easy to wire and pass inspection with ease.*

system performance, safety disconnects, and the charge controller. Power centers provide connection points to which the wires to the battery bank, the inverter and the wind generator connect. Although power centers may cost a bit more than buying all the components separately, they are easier and cheaper to install.

Pros and Cons of Off-Grid Systems

Off-grid systems provide freedom from power outages, energy independence, and total emancipation from the electric utility (Table 3.3). If designed and operated correctly, they will provide sufficient energy to meet your needs for many years.

Although, they do free you from utilities, you will still very likely need to buy a generator (gen-set) and fuel to power it. Gen-sets produce pollution and cost money to maintain and operate. Off-grid systems are also the most expensive of all systems because of the need for batteries and backup power (via PV systems and/or gen-sets), which add substantially to the cost. They also require more wiring and additional space to house battery banks and generators. They require more maintenance, too, thanks to the batteries and generators. Batteries require replacement every five to ten years, depending on the quality of batteries you buy and how well you maintain them. Battery production and recycling also exact a toll on the environment.

Table 3.3 Pros and Cons of Off-Grid Systems	
Pros	**Cons**
Provide a reliable source of electricity	Generally the most costly wind energy system
Provide freedom from the utility grid	Less efficient than batteryless grid-connected systems
Can be cheaper to install than grid-connected systems if located more than 0.2 miles from grid	Require more maintenance than batteryless grid-connected systems (you take on all of the utility's operation and maintenance jobs and costs)

Although cost is a major downside, there are times when off-grid systems cost the same or less than grid-connected systems — for example, if a home or business is located more than a few tenths of a mile from the utility lines. Under such circumstances, it can cost more to run electric lines to a home than to install an off-grid wind system.

Hybrid Systems

Many homeowners and business owners, especially in rural areas, install hybrid systems to meet their needs, as discussed earlier (Figure 3.8). In fact, most residential off-grid wind systems in use in North America are hybrids that combine solar electricity with wind.

In some locations (like the North American Great Plains) you can count on consistent winds for power production. But for most of the world, wind turbines and PVs are a marriage made in heaven because winds vary throughout the year. They tend to be strongest in the fall, winter and spring — from October or November through March or April. At Dan's educational center in central Missouri, The Evergreen Institute's Center for Renewable Energy and Green Building, for instance, the highest average wind speeds occur from October through May (Figure 3.9). During these

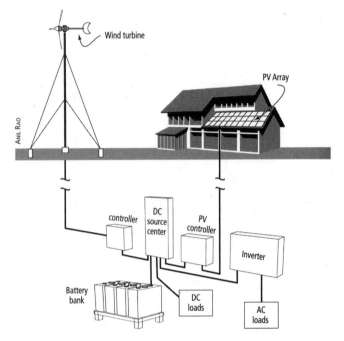

Fig. 3.8: *Hybrid System. This system consists of a PV array and wind turbine, two renewable energy technologies that complement each other very nicely.*

months, a properly sized wind generator can meet most of a family or business's needs.

Winds continue to blow, but less frequently and less forcefully through the rest of the year. Fortunately, though, sunshine is more abundant during this less-windy period. In such instances, a solar electric system can supplement a wind energy system, providing the bulk of the electricity while the wind turbine plays a backup role.

Because solar and wind resources are often complementary, hybrid systems provide a more consistent year-round output than either wind-only or PV-only systems. Sized correctly, in areas with a sufficient solar and wind resource, hybrid wind/PV systems can provide 100 percent of one's electricity This complementary relationship is shown graphically in Figure 3.9.

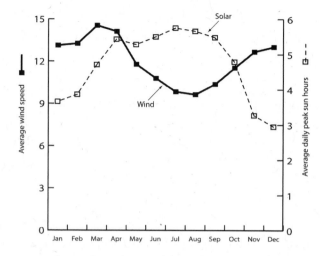

Fig. 3.9: *The Complementary Nature of Wind and Solar. Wind and solar energy often complement each other, creating a reliable, year-round source of electricity. Note how well solar makes up for the reduction in wind during the summer months at Dan's education center, The Evergreen Institute, in east-central Missouri.*

A hybrid wind/PV system may even eliminate the need for a backup generator. Moreover, hybrid systems often require smaller solar electric arrays and smaller wind generators than if either were the sole source of electricity. However, if the combined output of a solar and wind system is not sufficient throughout the year, you will either need to cut back on consumption or run a backup generator. A generator is useful because it is used to maintain batteries in peak condition (discussed in Chapter 7), and it may allow installation of a smaller battery bank.

Choosing a Wind Energy System

By far the cheapest and simplest wind energy option is a battery-less grid-connected system. The occasional power outage will shut your system down, but in most places these are rare and short-lived events. Grid-connected systems are usually cheaper than going off-grid. Moreover, if you are installing a wind system on an existing home that is already connected to the grid, it's best to stay connected. Use the grid as your batteries.

Grid-connected systems with battery banks are suitable for those who want to stay connected to the grid, but want protection against occasional blackouts or brownouts. They cost more, but provide peace of mind and security.

However, if you are building a new home and you are more than a few tenths of a mile from existing power lines, connecting to the grid can be expensive. While some utility companies foot the bill for line extension, others charge to run an electrical line to your home. Utility company policies vary considerably when it comes to line extension costs. Hook-up fees can run up to $50,000 — or more — if you live more than half a mile from the closest electric lines. Dan has a client who spent $65,000 to run a utility line two miles to her property, which could have purchased a huge renewable energy system for her cabins. So be sure to check with your local utility when considering which system you should install. In some locations, a quarter-mile grid connection is costly enough to justify an off-grid system.

CHAPTER 4

WIND SITE ASSESSMENT

Wind is an abundant and clean energy resource available in many locations throughout the world. Even though you may live on a windy site, but before you buy a system, it is a good idea to know if your site really has sufficient wind resources to merit the investment. If it does, a wind system could represent a wise financial decision.

We should point out, however, that a wind energy system could also make sense for other reasons. It may make sense for environmental reasons. It may also become a fun hobby that pays a little. Or, it may be your contribution to the ever-growing industry or to creating a better future for your children, the rest of humanity and the many species that share this planet with us.

As a rule, the decision to install a wind energy system is based on a combination of factors. In this chapter, we'll focus on economics of wind systems, never losing track of other reasons.

If the economics of a wind system is your primary focus, the decision to invest time and money on a system hinges on three factors: (1) your electrical energy requirements — how many kilowatt-hours of electricity you need, (2) the number of kilowatt-hours of electricity a wind turbine could produce at your site, and (3) the cost of supplying electricity from a wind system versus other sources, for example, the local utility, or other renewable energy technologies, such as solar electricity. As you shall see, several factors help to make small wind an economically viable — or even profitable — venture. They include: (1) high electricity rates, (2) rebates or tax credits from utilities or governments, (3) a good wind resource, and (4) a long-term perspective.

To begin, you first need to determine your electrical demand. This will tell you how much electricity you'll need to generate from your system.

Assessing Electrical Demand

Electrical energy consumption varies widely from one household or business to the next. In an all-electric home equipped with numerous electricity-guzzling appliances — like electric stoves, central air conditioners, electric space heating and electric hot water — average monthly electrical consumption typically falls within the 2,000 to 3,000 kilowatt-hour range. In homes that use natural gas or propane to cook, heat water, and provide space heat — and have no air conditioner — electrical consumption may be as low as 400 to 500 kilowatt-hours per month. On average, though, most homes in the United States consume between 800 and 1,000 kilowatt-hours of electricity per month.

How one goes about assessing the electrical consumption of a residence or business depends on whether it is an existing structure or one that's about to be built.

Assessing Demand in Existing Structures

Assessing the electrical consumption of an existing home or business is fairly easy. Simply review monthly electric bills, going back two to three years, if possible.

If you don't save electric bills, a telephone call to the utility company will usually yield the information you need. Most utilities will gladly provide data on energy use to their customers. Electrical consumption is listed as kilowatt-hours. Many companies also post the data online. All you need is your customer number.

If you recently purchased an existing home, you can ask the previous owners to share their utility bills. If they have not saved them, they may be willing to contact the utility to request the information. Remember, however, a house doesn't consume electricity, its occupants do, and we all use energy differently. If you are frugal and the previous owners were not, their past energy usage data may be of little value to you in predicting your usage.

Once you have secured energy bills, calculate the total annual electrical consumption. Do this by adding the monthly totals. If you have data from two or more years, calculate a yearly average. If several years' data are available, it is helpful to determine the range — that is, the lowest year and the highest year. It's important to look for trends, too. For example, is energy use on the rise, staying constant, or declining from year to year? If you notice a dramatic increase in electrical energy demand in recent years, exclude earlier data, which lowers the annual average. More recent data more closely reflects your electrical consumption. If, on the other hand, electrical energy consumption has declined, earlier data will inflate electrical demand.

If you are installing a grid-connected system, this is all the data you'll need. You or an installer can size your system based on the annual electrical consumption and the percentage of electricity you would like to generate by your wind turbine.

If you are installing an off-grid system, you need to calculate monthly averages — that is, how much electricity is used, on average, during each month of the year. To do this, simply add up household electrical consumption for each month, and then divide by the number of years' worth of data you have. If your records go back three years, add the electrical consumption for all three Januarys, then divide by three. Do the same for February and the remaining months. Record monthly averages on a table.

After you have calculated monthly averages, look for patterns in energy use. Are there months or seasons during which electrical demand is higher or lower than others? If you live in a hot, humid climate, for example, electrical consumption may peak in the summer when the air conditioner is operating full tilt. If you live in a colder climate in which air conditioning is not required, electrical consumption may peak in the winter because lights are on the longest and the heating system is being used. Understanding seasonal demand is extremely important when sizing an off-grid system because you need to size your system to meet your demand during that period.

Assessing Demand in New Structures

Estimating electrical consumption in a new building — one that's just been built or is about to be built — is more difficult, because here's no historical energy consumption data.

One way to estimate future electrical consumption in such instances is to base it on usage in your current home. If you are building a new home that's the same size as your current home and has the same amenities, electrical consumption may be similar to that in your current abode. However, if you are building a more energy-efficient home and installing much more energy-efficient lighting and appliances, electrical consumption could easily be 30 to 50 percent lower, perhaps even 75 percent below that in your current home. If that's the case, adjust current electrical consumption to reflect the savings you'll enjoy.

Another technique is a load analysis. To perform a load analysis, list all the appliances, lights and electronic devices you are installing in your new home or business. Then determine how much electricity each one uses and how many hours per day each one operates. From this data, you can approximate demand in kilowatt-hours.

Electrical load analysis is pretty straightforward. It's made even simpler by worksheets. You can find them online. Professional renewable energy installers also often supply customers with worksheets, and may even help customers fill them out.

Begin by listing all the electrical devices in your new home or business. Rather than list every light bulb separately, however, you may want to lump them together. Once you've completed the list, you need to determine the amount of electricity each item consumes. This can be determined by consulting charts that list typical wattages for common electrical devices. You can find detailed charts in various books, such as Solar Energy International's *Photovoltaics: Design and Installation Manual*, or John Schaeffer's *Solar Living Source Book*. You can also look at the website of WE Energies, a Wisconsin utility. It contains a chart that will help you determine the approximate electrical consumption of appliances and other devices. Their website is: webapps.we-energies.com/appliances/apply_calc.cfm.

Energy consumption in a new home can be more accurately determined by consulting the nameplate on the back or bottom of electrical devices. It typically lists the unit's power consumption in watts. It also typically lists amperage and voltage. Unfortunately, there's no universal standard for reporting this information. Voltage may be listed as 120 volts, 120 V, 120 volts AC, or 120 VAC. They are all the same. In some cases, the stickers only list voltage and current (amperage). To calculate watts, simply multiply the two (watts = amps x volts). This will provide an accurate value of watts for devices such as electric heaters and incandescent light bulbs. For motors and fluorescent light fixtures, amps x volts overestimates wattage. Multiply by a factor of about 0.7 to get a better estimate of watts for these devices.

Load analysis requires use of the "run wattage," that is, the maximum wattage an appliance draws when in operation. However, the run wattage of an appliance typically represents a worst case estimate — for example, the wattage of a TV at full volume. In such instances, you can reduce the nameplate wattage by about 25 percent (multiply wattage by 0.75).

Another way to determine wattage is to measure it with a watt meter like those shown in Figure 4.1. The meters are plugged into electrical outlets and electronic devices are then plugged directly into them. The meter indicates instantaneous power (watts).

After you have recorded the run wattage of each device, you need to estimate the number of hours each device is used each day. Multiplying the two yields watt-hours. You then need to determine the number of days each device is used during a typical week in your home or business. From this information, you calculate the weekly energy consumption of all your devices — lights, appliances,

DAN CHIRAS

ELECTRONIC EDUCATION DEVICES

a   b

Fig. 4.1: *The cleverly named Kill A Watt meter (a) and Watts Up? Meter (b) are valuable tools for measuring the wattage of electronic devices and also ferreting out phantom loads.*

stereos, tools, etc. — in watt-hours. You then divide this number by 7 to obtain the average daily consumption in watt-hours.

While straightforward, this system is tedious and subject to error. That's because it is difficult for many of us to accurately estimate how long appliances and lights run on a typical day. For example, how many hours do you operate your toaster each day? Is it three minutes, five minutes, or ten? And how often and how long do you run the blender? How long is the kitchen light on? How long does the refrigerator run?

Another problem is that electric consumption varies by season. Electrical lighting, for example, is used much less in the summer than the winter, because days are longer in the summer and people often spend more time outdoors. In addition, furnace blowers operate a lot during the winter, but very little in the late spring and early fall and not at all in the summer. So what's the average daily run time of the furnace blower?

Another problem with this approach is that many electronic devices draw power when they're off. These are known as *phantom loads*. Television sets, VCRs, satellite receivers, cell phone and laptop computer chargers, and a host of other devices all draw power after you've turned them off — sometimes nearly as much as when they're operating. Phantom loads typically account for five to ten percent of the monthly electrical consumption in US homes and, therefore, can add to the overall load, although they are rarely factored into a load analysis.

Because of these problems, homeowners often grossly underestimate their consumption of electricity. If you use this technique, be careful.

To avoid underestimating electrical demand, it's not a bad idea to compare projected energy use determined by a load analysis to the energy consumption in your existing home or business. If your estimate and actual energy usage differ considerably, and you can't explain why, you may want to start over.

Conservation and Efficiency First!

Once you've projected electrical energy use, it's a good time to consider implementing energy-efficiency measures — that is, finding ways to slash electrical demand. Energy efficiency measures help reduce load. The lower your energy demand, the smaller your wind-electric system will need to be. The smaller the system, the less money you'll have to put out initially.

The cheapest way to reduce electrical demand is by behavioral changes such as turning off lights and turning down thermostats in the winter. These are energy conservation measures and are the lowest-hanging fruit on the energy-saving tree. Although simple, they may be the most difficult to enact because they involve changing personal habits.

Huge amounts of energy can also be saved by weatherizing a home or business by installing weather stripping and caulking leaks, then beefing up the insulation in the roof and walls and under floors. Weatherization and insulation not only save money on heating oil and gas, they also cut electrical use in homes heated by fossil fuels by reducing the run time of furnaces and boilers. In the summer, these measures reduce the run time of ceiling fans, air conditioners and evaporative coolers, saving even more energy.

The next most cost effective way to cut demand is through energy efficient technologies, such as energy-efficient lights, washing

Fig. 4.2:

Energy-Efficient Washing Machine. Energy-efficient appliances like this Frigidaire Gallery horizontal axis (front-loading) washing machine decrease the consumption of electricity and offer the best return on investment. When contemplating a renewable energy system, be sure to make your home as energy efficient as possible first.

machines, dishwashers, furnaces, air conditioners and new windows (Figure 4.2). Remember, though, that while all of these are vital to creating a more energy-efficient way of life or business, they're the highest fruit on the energy-efficiency tree — and thus the most expensive.

One of the biggest big-ticket items is the refrigerator. Refrigerators may be responsible for a staggering 25 percent of electrical consumption in a home not heated or cooled electrically. If your refrigerator is more than ten years old and is in need of replacement, recycle it, and buy a new energy-efficient model. Whatever you do, don't lug the old fridge out to the garage or take it down to the basement and use it to store an occasional case of beer. It will undermine your efforts to create an energy efficient household and rob you blind!

Waste can also be reduced by installing Energy Star qualified televisions and stereo equipment (Figure 4.3). And, of course, you can trim more fat from your energy bill by installing energy-efficient lighting, such as compact fluorescent lights.

Although efficiency has been the mantra of the energy advocates for many years, don't discount its importance just because the advice is a bit threadbare. Few people, it turns out, have heeded the

Fig. 4.3:
Energy Efficient Stereo with Energy Star Label. This label assures you that the product you're looking at is one of the most energy efficient in its category.

calls for energy efficiency. And many who have may not have fully tapped the potential savings.

Those interested in learning more about making their homes energy efficient may want to read the chapters on energy conservation in Dan's book, *Green Home Improvement*, which describes numerous projects to reduce energy use in homes and businesses.

For an up-to-date list of energy-efficient appliances, US readers can visit the EPA and DOE's Energy Star website at energystar.gov. Click on appliances. Consumer Reports has an excellent online list of energy-efficient appliances, and it also rates them on reliability. Canadian readers can log on to oee.nrcan.gc.ca/energystar/english/consumers/index.cfm for a list of international Energy Star appliances.

Assessing Your Wind Resource

Once you have executed a strategy to make your home or business — and its occupants — more energy efficient and have recalculated energy consumption, it is time to determine how much wind is available to you and when it is available. From this information, you can select a wind turbine that will produce enough electricity to meet your needs.

An assessment of available wind resources can be made using one or a combination of techniques. Professional small wind site assessors, however, rely principally on state wind maps and online sources.

State Wind Maps

One of the best and easiest ways to assess the wind resources in a region is to consult a state wind map. State wind maps list average annual wind speed in meters per second and miles per hour. Unfortunately, wind speed estimates on the map are reported at 165 and 195 feet (50 or 60 meters) above the surface of the ground[2].

Most states have good wind maps. You can locate your state's wind map at the Wind Powering America website of the DOE at eere.energy.gov/windandhydro/windpoweringamerica/index.asp. It's probably faster to search for "Wind Powering America" than to type out this URL. Another great source is www.awstruewind.com. AWS Truewind prepared most of the Wind Powering America maps. The *Canadian Wind Atlas* can be obtained by visiting cmc.ec.gc.ca/science/rpn/modcom/eole/CanadianAtlas.html.

While wind maps are an excellent source of information for your area, they do have some limitations. One of them is resolution. In some areas, like the plains of western Kansas, wind maps show uniform wind speeds over large areas. If you live in one of those areas, the map will give you a pretty accurate idea of average wind speed. However, wind speeds over more complex topography can vary considerably over short distances. The resolution of the maps isn't good enough to pinpoint an exact location. Dan's home in the foothills of the Rocky Mountains is a good example. The state wind map shows three different classes of winds in his area. Which one applies depends on the specific location. A neighbor a tenth of a mile away in a valley may have very little wind, while a neighbor perched on top of a mountain will have a great deal of wind. In general, the more complex the terrain, the less accurate the wind maps are.

---

2. State wind maps contain estimates, not direct measurements of wind speed. Direct measurements are used to validate the maps, but maps are created using upper air wind speed data (wind balloon data) and topographic data. Computer models then predict the wind speed at 50 meters (165 feet) for the entire state. Most residential wind machines, however, are mounted at about 80 to 120 feet. A wind site assessor uses a simple equation to extrapolate downward.

When assessing wind resources, remember that hills, cliffs, forests and buildings can reduce wind speed. However, some types of hills and cliffs can magnify winds, as shown in Figure 4.4. It depends on where you're located. In most cases, though, wind maps are adequate for siting a small wind turbine. That said, it's always a good idea to assess the topography, vegetative cover, and ground clutter when estimating average annual wind speed. If you live in an open field on top of a knoll, your wind resource may be significantly higher than a neighbor who lives a mile away in a tree-covered valley.

Online Databanks

While state wind maps are an excellent source of information, a few states have not invested in the best mapping technology. Because of this, some professional small wind site assessors perform small wind site analyses using a very extensive online database developed by NASA. The website, Surface Meteorology and Solar Energy, is at eosweb.larc.nasa.gov/sse.

This site provides a wealth of data on wind energy, including tables that show both the monthly and annual average wind direction and average wind speed at sites throughout the world. The site is pretty user friendly, but you will need to start by setting up a free account.

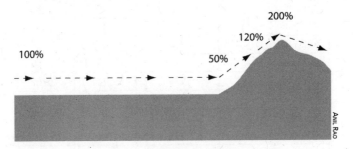

Fig. 4.4: *Effect of Topography on Wind Speed. Hills can dramatically increase wind speed. Placement of a tower on the top of the hill could result in a significant increase in the power output of a turbine. Placement of a turbine at the base of a hill could result in much lower output.*

Like wind maps, NASA data provides good information, but you should still assess your site very carefully, especially in complex topography. If your site is in a valley or surrounded by trees or buildings, the wind speed at the hub height may be significantly lower than the satellite data suggests.

When in doubt, hire a professional wind site assessor to analyze your site and make recommendations for tower/turbine placement and minimum acceptable tower height. A list of certified wind site assessors is posted online at the Midwest Renewable Energy Association's website, the-mrea.org. They're the national certifying body for wind site assessors. You can also ask a professional installer to render an opinion, although not all installers are as knowledgeable about wind and wind site assessment as certified wind site

Table 4.1
kWh/year Outputs of Representative
Wind Turbines (2-13-08)

wind in mph	swept area	8	9
Whisper 100	39	360	540
xl 1	53	660	1,020
WT 600	55	504	792
Whisper 200	63.5	720	1,080
WT 2500	97	2,004	2,472
ARE 110	110	1,620	2,316
Skystream	113	1,200	2,040
Whisper 500	176	2,040	2,760
Endurance	254	1,843	3,091
WT 6000	254	5,004	6,768
BWC XL-S	415	2,880	4,440
ARE 442	442	7,476	10,440
Jacobs 31-20	754	9,828	13,920
V-15-35	1,964	N/A	38,000
PGE 20/32	3,120	N/A	53,280

MICK SAGRILLO

assessors. Some installers may be more interested in selling and installing a turbine and tower than in giving you an objective wind site assessment.

How Much Electricity Will a Wind Turbine Produce?

Once you've determined the average wind speed at a site, it is time to determine how much electricity a wind generator could produce at the proposed tower height — and therefore whether it can meet all your needs or what percentage of your needs it will satisfy. This step is fairly easy.

Table 4.1 shows a list of wind turbines and the estimated annual output of each turbine (in kilowatt-hours) at seven different average wind speeds. To see how this table is used, consider an example.

10	11	12	13
780	960	1,200	1,500
1,380	1,800	2,256	2,640
996	1,356	1,488	1,752
1,500	1,920	2,280	2,700
3,516	3,996	5,004	5,580
3,144	4,068	5,040	6,060
2,880	3,720	4,560	5,400
3,960	4,920	6,456	7,440
4,587	6,268	8,068	9,920
8,004	11,004	12,996	15,000
6,240	8,400	10,800	13,560
14,052	17,640	21,972	25,584
19,728	25,704	32,292	39,288
43,000	58,000	64,000	80,000
64,920	82,296	90,000	107,796

Let's assume that the average wind speed at a site (at hub height) is 12 miles per hour. Let's also assume that your load analysis, after efficiency measures have been implemented, indicates you'll need, on average, 900 kilowatt-hours per month, or 10,800 per year. In the 12 mile-per-hour column, you'll discover two wind turbines that match your electrical requirements, the WT6000 (by Proven) and BWC (Bergey Wind Power's) XL-S. If the wind speed at your site is 13 miles per hour, an Endurance wind turbine would meet your needs. The Proven WT 6000 and the Bergey XL-S would produce more than you need.

Annual energy outputs used to estimate the economic performance of a wind energy system can also be found in an article Mick and Ian published in *Home Power* magazine (Issue 131), entitled "How to Buy a Wind Generator."

You can also obtain annual energy output data directly from wind turbine manufacturers. While this data is useful, manufacturers tend to overstate the electric production of their turbines. As a result, we recommend derating their estimated outputs by 20 percent — just to be conservative. (Note that the data from Wisconsin's Focus on Energy *is* derated annual energy output; the data in Ian and Mick's article is manufacturer data and is *not* derated.)

Once you've found the wind turbine that meets your needs, you need to be sure it is the appropriate type of turbine. That is, you need to be sure you select a grid-tied turbine or battery-based turbine, depending on the type of system you are planning on installing.

Does a Wind System Make Economic Sense?

At least three options are available when it comes to analyzing the economic cost and benefits of a wind turbine: (1) a comparison of the cost of electricity from the wind turbine with conventional power or some other renewable energy technology, (2) an estimate of return on investment, and (3) a more sophisticated economic analysis tool known as discounting. We'll present an overview of each method in this chapter.

Cost of Electricity Comparison

One of the simplest ways of analyzing the economic performance of a wind energy system is to compare the cost of electricity from the wind system to the cost of electricity from a conventional source — notably, the local utility — or some other renewable energy technology you are considering.

To begin, you must determine the annual electrical consumption of your home or business. Second, determine the average monthly wind speeds at the site. Third, identify a wind turbine that produces a sufficient amount of electricity to meet your needs. If you are contemplating an off-grid system, you want to size the system to meet your needs during the period of highest demand. If you're installing a utility-tied system, you only need an annual match.

In the example presented earlier, a Proven WT6000 would produce 12,996 kilowatt-hours per year. Now multiply the kilowatt-hours produced in a year by the life of the system. How long would that be?

A well-made, heavy-duty wind turbine like the Proven WT 6000 could last 20 to 30 years, with regular inspection once or twice a year and maintenance and repairs as required. A lighter-weight and cheaper model might only last five years or fewer. If the Proven WT6000 lasts 30 years, it would produce about 389,880 kilowatt-hours over its lifetime. In 2008, the Proven installed on a 120-foot guyed tower cost a little over $60,000 in the United States. Dividing the cost of the system by the total output yields the cost of electricity per kilowatt-hour. In this example, then, the electricity, over the lifetime of the turbine, will cost about 15 cents per kilowatt-hour.

Now it is time to compare the cost of electricity generated by the wind turbine to electricity from your utility — or possibly some other renewable energy source. When calculating the cost per kilowatt of utility power, be sure to add in all the costs — that is, taxes, fees and fuel surcharges the utility includes. You don't need to add the meter reading fee if you are installing a grid-connected system, as you'll be paying this fee if you buy from the utility or generate your own electricity. If the alternative source costs 15 cents per kilowatt-hour or more, the Proven would be a pretty good investment. Although the turbine will require maintenance and repair over the years, the cost of electricity from the utility is also bound to increase. It's been rising, nationwide, at a rate just under 4.5 percent annually for the last 35 years and could increase more rapidly as energy prices rise. It is likely, then, that maintenance costs and rising costs could offset each other.

This system would make economic sense if you were paying 15 cents per kilowatt-hour or more for locally generated electricity. That is, if you don't mind prepaying your electrical bill by laying down $60,000 all at once. You'll either need to withdraw $60,000 from a savings account or some other investment or take out a loan.

When calculating the cost of electricity from a wind system, don't forget to subtract financial incentives from federal, state and local governments, and local utilities. These incentives can be substantial. The federal incentive for wind, for example, is currently 30% of the system cost. Several states also offer incentives, either through state government or via local utilities, including New York, New Jersey, Wisconsin, Massachusetts, California and Oregon. In addition, the US Department of Agriculture also offers a 25 percent grant to cover the cost of wind systems on farms and rural businesses. To learn more about incentives in your state, look at the Database of State Incentives for Renewables and Efficiency at dsireusa.org.

If you are building a new home, don't forget to include the cost of connecting to the electrical grid when comparing the cost of wind-generated electricity to the cost of power from the utility.

Calculating Simple Return on Investment

Another relatively simple way of determining the cost effectiveness of a renewable energy system is to determine the simple return on investment (ROI). Return on investment is the rate of return expressed as a percentage of an investment.

ROI can be calculated by dividing the annual value of electricity generated by a wind system by the cost of the system. Let's calculate the return on investment for the $60,000 Proven WT6000 used previously. If the turbine costs $60,000 installed and produces 12,996 kilowatt-hours per year and electricity from the utility costs 15 cents per kilowatt-hour, the electricity would be worth $1,950 per year. To calculate the return on investment, divide the annual value of electricity by the cost of the system ($1,950 divided by $60,000). In this instance, the return on investment would be 3.2 percent. Not terribly good, but not bad either, especially given the state of the world economy and once you factor in the feel-good variables, like producing your own power from a clean, renewable energy source.

If this system were installed with a 30 percent tax credit from the federal government, the simple annual return on investment would be 4.3 percent ($1,950 divided by $45,000), which is much better than a certificate of deposit or even a mutual fund (at this writing). If a USDA grant were also obtained, the cost of the system would be reduced even more, increasing the return on investment.

Like the previous method, simple return on investment is just that, a crude way of estimating the economic performance of an investment. Neither method takes into account a number of other economic factors such as: (1) interest payments on loans required to purchase the system or lost interest if the system is paid from cash withdrawn from an investment, (2) insurance costs, or (3) property taxes. All of these lower the ROI.

This method also fails to factor in the increase in the cost of electricity from the local utility, which would make wind-generated electricity more valuable. Nor does it take into account possible income tax benefits for businesses, for example, accelerated depreciation. Return on investment also doesn't take into account the fact that the system depreciates in value over time as it ages.

Despite these shortcomings, simple return on investment is a nice way to evaluate the economic performance of a renewable energy system. It's light years ahead of the black sheep of the economic tools, payback.

Payback is a term that gained popularity in the 1970s. It is used for energy conservation measures and renewable energy systems. Payback is the length of time it takes a system or energy conservation measure to pay back its cost through the savings.

For a wind energy system, payback can be determined by dividing the cost of the system by the anticipated annual savings. If the $60,000 wind energy system with a 30% federal tax credit produces 12,996 kilowatt-hours per year and grid power costs you 15 cents per kilowatt-hour, the annual savings of $1,950 yield a payback of 23 years ($45,000 divided by $1,950). In other words, it will take the savings from your system 23 years to pay off the cost, ignoring the maintenance and repair costs. From that point on, the system produces electricity free of charge.

Simple payback has very serious drawbacks. The most important is that it can be misleading, as this example clearly illustrates: a 23-year payback seems ridiculously long, while 4.3 percent ROI seems pretty good — but they're two ways of looking at the very same investment!

Simple payback is also a concept we rarely use. Do anglers calculate the payback on their new bass boats? Do couples calculate the payback on the new chandelier?

Simple payback and simple return on investment are closely related metrics. Mathematically speaking, return on investment is the reciprocal of payback. That is, $ROI = 1/payback$. For example, a wind system with a 10-year payback represents a 10 percent return on investment ($ROI = 1/10$).

Discounting and Net Present Value: Comparing Discounted Costs

For those who want a more sophisticated tool, economists have developed a technique known as *discounting*. Unlike the previously discussed methods, discounting factors in numerous economic factors such as the maintenance costs, the rising cost of grid power, and another key element, the time value of money.

The time value of money takes into account the fact that a dollar today is worth less than a dollar tomorrow and even less than a dollar a few years from now. Economists calculate the loss of value by applying a discount factor. The discount factor represents something economists refer to as opportunity cost and it includes inflation. Opportunity cost is the cost of lost economic opportunities by pursuing one investment path over another — for example, investing your money in a solar system instead of in the stock market.

To make life easier, this economic analysis can be performed by using a spreadsheet like the one shown in Table 4.2. The first column is the year. The second column includes the discount factor. For simplicity, we recommend choosing the highest interest rate on any debt you have, including your mortgage, as the discount factor. Or, if you have no debt, choose the highest investment interest rate you can get with a risk profile similar to the renewable energy system, which is usually very low. A ten-year government bond is a good basis and currently pays less than 3%. As indicated in column 2, a dollar today will be worth 55 cents in 20 years.

The next column (under the category "Buy Utility Electricity") shows the cost of electricity from the local power company — that is, how much you will pay each year for electricity if you purchased it from the local utility rather than generating with a wind turbine. This column factors in the rising cost of electricity using a 4.4% annual increase. As shown here, the wind system produces $1,950 worth of electricity in year one. In year 2, that electricity would cost you $2,036 because of the rising cost. The last entry in column 3 is the total cost of electricity to you — $60,537. That's how much money you will pay the utility over the 20-year period if you purchase 12,996 kWh of electricity per year from them (with an inflationary increase of 4.4% per annum).

The next column under the heading "Buy Utility Electricity" is the discounted cost of electricity from the utility. The discounted cost of electricity from the utility is the cost of electricity taking into account the discount rate (the declining value of the dollar due to inflation) applied to the rising cost of electricity. This calculation allows one to calculate "present value" of the money spent on electricity from the utility over a 20-year period. Put another way,

Table 4.2
Analysis of a Wind System vs. Electricity from the Utility

Year	Discount Factor	Buy Utility Electricity	
Rate:	3.0%	Cost 4.4%	Discounted Cost
0	1.000	$0	$0
1	0.971	$1,950	$1,893
2	0.943	$2,036	$1,919
3	0.915	$2,125	$1,945
4	0.888	$2,219	$1,971
5	0.863	$2,317	$1,998
6	0.837	$2,418	$2,025
7	0.813	$2,525	$2,053
8	0.789	$2,636	$2,081
9	0.766	$2,752	$2,109
10	0.744	$2,873	$2,138
11	0.722	$2,999	$2,167
12	0.701	$3,131	$2,196
13	0.681	$3,269	$2,226
14	0.661	$3,413	$2,256
15	0.642	$3,563	$2,287
16	0.623	$3,720	$2,318
17	0.605	$3,884	$2,350
18	0.587	$4,055	$2,382
19	0.570	$4,233	$2,414
20	0.554	$4,419	$2,447
Total		$60,537	$43,176

that's the value of the money one would spend over a 20-year period in present-day dollars.

As you can see, although you will have shelled out $60,537 to the utility company, the net present value of that money — that is, the

| Proposed Wind System ||
Cost	Discounted Cost
$45,000	$45,000
$0	$0
$200	$189
$0	$0
$200	$178
$0	$0
$200	$167
$0	$0
$200	$158
$0	$0
$200	$149
$0	$0
$200	$140
$0	$0
$200	$132
$0	$0
$200	$125
$0	$0
$200	$117
$0	$0
$200	$111
$47,000	$46,466

$60,537 you will spend on electricity in 20 years is only worth $43,176 in present dollars.

In the fifth column of the spreadsheet is the cost of the wind energy system — $60,000 minus the 30% federal tax incentive or $45,000. Note that $200 is added every other year for maintenance. Over a period of 20 years, you will have invested $47,000 in your system (in present dollars).

The last column of the spreadsheet is the discounted cost of the wind system. This is the present value of your expenditure, taking into account the discount factor of 3%.

The final step is to compare the discounted cost of the system ($46,466) to the discounted cost of electricity from the utility ($43,176). In this example, the present value of the wind system is $3,290 more than the present value of the cost of utility electricity.

In this technique, if a present value of a wind system is lower than the present value of buying electricity, it makes economic sense. If it costs more, it doesn't. The greater the difference in the cost of the two systems, the more compelling the decision. Even if the differential is small, however, the investment may be worth it.

Putting It All Together

Comparing wind systems against the "competition," calculating the return on investment, or comparing strategies based on net present value are all ways to judge the economic feasibility of a wind energy system. As noted, economics is not the only measure on which we base our decisions. Energy independence, environmental values, reliability, the "cool" and "fun" factors, and the value of a hobby for renewable energy gear heads, among other factors, also figure prominently in our decisions to invest in wind. They are all perfectly valid motivations and should not be forgotten.

Economics may be irrelevant if you are committed to creating a sustainable future, helping reverse costly global climate change, creating a better world for future generations, setting a good example, or supporting the fledgling wind energy industry. All are perfectly acceptable reasons for installing a system.

CHAPTER 5

A Primer on Wind Generators

Your choice of wind turbines is somewhat limited. Even so, you need to know your options and choose wisely. This chapter will help you understand your options. We'll begin with the basics by examining the types of wind generators and then explore the components of the most common turbines: the horizontal axis machines. We'll also briefly examine vertical axis wind turbines and dispel the common myths about them. We'll also discuss features that will ensure many years of low-maintenance. We follow this discussion with a brief exploration of homemade wind machines.

The Anatomy of a Wind Turbine

Most wind turbines on the market today are known as horizontal axis wind turbines or HAWTs (Figure 3.1a). The second type is the vertical axis wind turbine or VAWT (Figure 5.1b).

Horizontal Axis Wind Turbines

Horizontal axis wind turbines come in two basic varieties: upwind or downwind (Figure 5.2a and b). In most HAWTs, the rotor is located on the upwind side of the tower when the turbine is operating, hence the name, upwind turbine. Those in which the rotor is located on the downwind side of the turbine when the wind is blowing are referred to as downwind turbines.

Fig. 5.1: *(a) Horizontal axis wind turbine (b) Vertical axis wind turbine. Although there's a lot of interest these days in vertical axis wind turbines, they are mounted at ground level, which exposes them to unproductive low-speed winds.*

Fig. 5.2: *(a) Upwind turbine (b) Downwind turbine*

Upwind HAWTs consist of three main parts: (1) a rotor, (2) an alternator, and (3) a tail (Figure 5.3). As noted in Chapter 3, the rotor consists of blades attached to a central hub covered by a nose cone to improve aerodynamics. Most upwind turbines have three blades. This entire assembly rotates when wind blows past the blades, hence the name "rotor."

In many small wind turbines the rotor is attached to a shaft that's attached to an alternator, a device that produces AC electricity. As

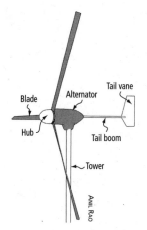

Fig. 5.3: *Anatomy of a Wind Generator. A wind turbine consists of blades attached to a hub, forming the rotor. The rotor in this turbine is attached via a shaft (not shown) to the rotor of the alternator. When the rotor of the turbine spins, it produces electricity.*

the name implies, in HAWTs the axis of the rotor is oriented horizontally, that is, parallel to the ground.

Alternators produce electricity. They consist of two main parts: a set of stationary windings, known as the stator; and a set of rotating magnets, known as the rotor. Most small wind turbines use metal magnets, rather than electromagnets. The movement of the magnets around the windings (as the blades spin) induces an electrical current in the windings. Wind turbines, therefore, first convert the kinetic energy of the wind into mechanical energy (rotation). The mechanical energy is then converted into electrical energy in the alternator.

HAWTs can also be classified according to their end use. Turbines designed to charge batteries are called "battery-charging turbines." Turbines designed to connect to the grid are referred to as "batteryless grid-tie turbines."

Figure 5.4 shows another common wind turbine configuration. In this design, there's no shaft. The blades of the turbine attach to a faceplate that is attached directly to a cylindrical metal "can." Together the blades and the face plate form the rotor of the turbine. The can to which the faceplate is attached contains magnets. They spin around a set of stationary coils of copper wire, the windings. The can is therefore also the rotor of the alternator. The windings constitute the stator of the alternator. As the rotor of the

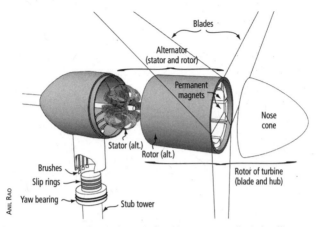

Fig. 5.4: *Anatomy of a Modern Wind Turbine. Many small wind turbines now directly link the rotor of the turbine to the rotor of the alternator, as shown here. The magnets are attached to the inside of the can, shown in the drawing. They rotate around the stationary windings of the alternator stator.*

turbine spins, the magnets rotate around the windings, producing AC electricity in the windings via electromagnetic induction.

Alternators in many modern small wind turbines contain high-strength rare earth magnets. Rare earth magnets contain neodymium, iron and boron. They produce a much stronger magnetic field than conventional iron or ceramic magnets used in some small wind turbines. The stronger the magnetic field, the greater the output of a wind turbine, all things being equal. Alternators that contain magnets such as these are referred to as permanent magnetic generators.

Electricity leaves the alternator via wires that attach to the stator. In most turbines, like the one shown in Figure 5.4, these wires terminate on metal brushes. The brushes, in turn, contact slip rings, which are brass rings located near the yaw bearing. The yaw bearing allows the turbine to turn in response to changes in the wind direction. The brushes transfer electricity from the alternator to the slip rings. The slip rings, in turn, connect to a cable (electric wire) that runs down the length of the tower. Electricity flows from the alternator to the brushes to the slip rings and then down the tower.

Alternator vs. Generator

A generator is a machine that converts mechanical energy into electrical energy, either AC or DC. A generator that produces AC is called an AC generator or, more commonly, an alternator. Alternators that produce DC electricity are known as DC generators or simply generators. Most, if not all, modern wind turbines contain alternators, which produce AC electricity. Some of these turbines are equipped with rectifiers, devices that convert the AC to DC, which is then sent down the tower.

Wind generators such as the ones shown in Figure 5.4 are known as direct drive turbines because the rotor of the turbine is attached directly to the rotor of the alternator. As a result, the rotor and the alternator turn at the same speed. Although virtually all modern residential wind machines are direct drive, a few of the small wind turbines contain gearboxes. They're located between the rotor of the turbine and the rotor of the alternator. Gearboxes increase the speed at which the alternator spins, increasing the output of the alternator. This allows the alternator to be much smaller and also maintains a rotor speed (blade speed) that is safe and quiet. These turbines are known as gear-driven turbines.

Another important component of most horizontal axis wind machines is the tail. The tail typically consists of a boom and a vane. The tail boom connects the tail vane to the body of the turbine.

Tails keep the rotor of the wind turbine pointing into the wind. If the wind direction shifts, the tale vane turns the turbine into the wind, ensuring maximum electrical energy production. The rotation of a wind machine on a tower as it tracks the wind direction is referred to as *yawing*.

Although upwind turbines dominate the market, several manufacturers produce downwind turbines: Proven, Southwest Wind Power (Skystream), Entegrity, Ventera and PGE. These wind turbines contain no tails.

Downwind turbines work well. However, if the wind dies down and then reverses direction, downwind turbines can get caught in the upwind position (with blades upwind from the tower). When stuck upwind, downwind turbines are unable to spin and generate electricity. As the wind speed increases or shifts direction, however, the turbine aligns properly.

Vertical Axis Wind Turbines

There's another type of wind machine that is getting a lot of attention these days. It is known as a vertical axis wind turbine. As shown in Figure 5.1b, the blades of a vertical axis wind turbine (VAWT) are attached to a central vertical shaft. When the blades spin, the shaft spins. The shaft is attached to an alternator generally located at the bottom of the shaft, often at ground level.

Vertical axis wind energy devices have been around for a long time, about 3,000 years. Proponents of VAWTs tout a number of supposed advantages over HAWTs, most of which are either wrong or grossly exaggerated. One of them is that they can capture wind from any direction. The machines don't need to be oriented into the wind as the HAWTs do. In addition, proponents like to claim that VAWTs are immune to turbulence that wrecks havoc with HAWTs.

Another supposed advantage is that VAWTs can be mounted close to the ground — even on top of buildings — where they capture ground-level winds. This eliminates the need for tall and costly towers and the need to obtain the zoning variances sometimes required to install horizontal axis wind turbines on tall towers.

Yet another supposed advantage of VAWTs stems from the fact that the generator can be mounted at ground level. This, say proponents, makes it easier to access and repair the generator should the need arise. There's no need to climb the tower — or lower a wind generator to the ground — to perform routine maintenance or to replace damaged parts.

Unfortunately, years of experience with VAWTS have been rather discouraging. "Hundreds of commercial VAWTs were installed in California in the late 1980s and early 1990s," Bob Aram reminds us. "They all failed and were removed from service. These were not experimental units, but production units." In addition, VAWTs are less efficient than horizontal axis wind machines. "For a given swept area," Jim Green notes, "they just don't extract as much wind energy as a well-designed HAWT." Moreover, the blades of VAWTs are prone to fatigue created as the blades spin around the central axis. The vertically oriented blades used in some early models, for instance, twisted and bent as they rotated in the wind. This caused the blades to flex and crack. Over time, the blades broke apart, sometimes leading to catastrophic failure. Because of these problems, VAWTs have proven less reliable than HAWTs.

"The VAWT does have an advantage in dealing with wind direction shifts," agrees Robert Preus, wind energy expert and manufacturer of Abundant Renewable Energy turbines (horizontal axis wind turbines). However, rapidly changing wind direction that occurs in turbulent low-level winds increases fatigue on a VAWT, just like a HAWT. "Fatigue leads to equipment failure, which has been a major problem with VAWTs."

Many VAWTs also require large bearings at the top of the tower to permit rotation of the shaft. When the top bearings or the blades need replacement, you've got a job on your hands.

Although VAWTs can capture ground-level winds, just like any turbine installed on a too-short tower, they are just as sensitive to turbulence and ground drag as horizontal axis wind turbines. As you learned in Chapter 2, ground-level winds are subject to friction (which creates ground drag). Both ground drag and turbulence in lower-level winds diminish the power available to any turbine mounted close to the ground — so much so that there is very little extractable energy in wind in such locations. The lower the wind speed, the less electricity a turbine will produce. In addition, dead air spaces form behind buildings and other ground clutter. Placing a VAWT in such a location renders it useless. So just because a VAWT can be mounted at ground level doesn't mean it will produce enough electricity to be worthwhile.

VAWTs are less reliable and less efficient than HAWTs. They just don't stack up against horizontal axis wind turbines. The few advantages they offer cannot counter the many, some say fatal, disadvantages.

The Main Components of Wind Turbines

To help you select a wind turbine that's built to last, let's take a closer look at the main components of modern wind turbines, starting with blades and generators.

Blades

Blades are a vital component of a wind turbine. Manufacturers make blades from several different types of materials. Although wood was once commonly used to make blades for small wind turbines, we're not aware of any companies that manufacture blades exclusively from wood. It's just too expensive and time consuming. The blades of modern wind turbines are typically made of extremely durable and relatively inexpensive synthetic materials, various types of plastic or composites — plastic reinforced with fiberglass or carbon fibers, for example. These synthetic blades typically last between 10 and 20 years. And, unlike the metal blades once used in small wind turbines, plastic blades don't interfere with television, satellite TV or wireless Internet signals. Table 5.1 lists some examples of blade materials commonly used today.

Table 5.1 Composition of Small Wind Turbine Blades	
Company	**Blade Material**
Abundant Renewable Energy	Fiberglass
Bergey Windpower	Fiberglass
Wind Turbine Industries	Fiberglass
Proven	Polypropylene (WT600) Fiberglass (WT2500, WT6000 and WT15000)
Southwest Windpower	Polypropylene reinforced with fiberglass Fiberglass

Most modern residential wind machines — both upwind and downwind models — have three blades. Although certain manufacturers produce two- and six-blade turbines (for sailboats), three-blade wind turbines are the industry standard. They provide the best overall performance. They are quieter and suffer less wear and tear than one- and two-blade models.

Generators

Most residential wind machines on the market contain permanent magnet alternators, described earlier, although larger turbines in the small turbine range, like the remanufactured Jacobs 31-20, incorporate electromagnets (magnets created by running electricity through wire coils in the alternator). As noted in Chapter 3, most wind turbines produce wild three-phase AC, initially. The wild AC output is converted (rectified) to DC electricity by a rectifier. The rectifier may be located in the wind turbine itself or in the controller. Controllers are typically mounted next to the inverter or may be incorporated in the inverter. DC electricity is then sent to the battery bank.

At least one manufacturer (Endurance) is now producing a small turbine that contains an induction generator. This wind turbine produces grid-compatible AC power without the use of an inverter. An induction generator looks a lot like electric induction motors used in modern society. When spun faster than its normal operating speed, an induction generator produces AC electricity. Because an induction generator synchronizes with the grid — that is, produces electricity at the same frequency and voltage — no inverter is required. But what features should you look for?

What to Look for When Buying a Wind Machine

While there are many turbines on the market, careful load and site analysis will narrow the field considerably. Once you have determined your average monthly electrical load and the average wind speed on your site, you can select a wind turbine that will produce enough electricity to meet your demands.

Manufacturers provide a plethora of technical data on their wind machines that can be used to make comparisons. Unfortunately, most of it is useless. Further complicating matters, "There can be a big difference in reliability, ruggedness, and life expectancy from one brand to the next," according to Mike Bergey, president of Bergey Windpower.

So how do you go about selecting a wind machine?

Although wind turbines can be compared using many criteria, there are only a handful that really matter: (1) swept area, (2) durability, (3) annual energy output, (4) governing mechanism, (5) shut-down mechanism, and (6) sound.

Swept Area

Swept area is the area of the circle described by the spinning blades of a turbine. Because the blades of a wind turbine convert wind energy into electrical energy, the swept area is the collector area of the turbine. The greater the swept area, the greater the collector area. The bigger the swept area, the more energy you'll be able to capture from the wind. To get the most out of a wind turbine — to produce the most electricity at the lowest cost — select a wind turbine with the greatest swept area. Swept area allows for easy comparison of different models.

Swept area is determined by rotor diameter. The rotor diameter is the distance from one side of the circle created by the spinning blades to a point on the opposite side or about twice the length of the blades. When comparing wind turbines, then, the rotor diameter is a pretty good measure of how much electricity a turbine will generate. Although other features such as the efficiency of the generator and the design of the blades influence energy production, for most turbines they pale in comparison to the influence of rotor diameter and, hence, swept area.

Manufacturers list the rotor diameter in feet or meters — often both. The greater the blade length, the greater the rotor diameter and the greater the swept area.

Most manufacturers also list the swept area of the rotor. Swept area is presented in square feet or square meters — sometimes both.

Annual Energy Output

Another, even more useful, measure is the annual energy output (AEO) or annual energy production (AEP) at various wind speeds. The AEO of a given wind turbine is presented as kilowatt-hours of electricity produced at various average wind speeds. Like the US EPA's estimated gas mileage for vehicles, AEO gives buyers a convenient way to compare models. As in the estimated gas mileage rating, however, AEOs won't tell you exactly how much electricity a wind machine will produce at a site. Performance varies depending on a number of factors such as turbulence and the density of the air.

Durability: Tower Top Weight

Another extremely important criterion is durability. The most important measure of durability is tower top weight — how much a wind turbine weighs. Four turbines that produce about the same amount of electricity are for example, the Proven WT2500 (419 pounds), the ARE110 (315 pounds), the Skystream 3.7 (170 pounds) and the Whisper 500 (155 pounds). The weight differences are in some cases substantial.

In our experience, heavyweight wind turbines tend to survive the longest — sometimes many years longer than medium or lightweight turbines. Weight is usually reflected in the price. Remember, however, that you get what you pay for. Producing electricity on a precarious perch 80 to 165 feet above the ground isn't a job you want to relegate to the lowest bidder, which is invariably the lightest turbine.

Balance of System Cost

Before you buy a machine, consider the total system cost. You'll need to purchase a tower and pay for installation, unless, of course, you install the tower yourself. Even then, you'll need to pay for concrete, rebar and equipment to excavate the foundation and anchors. You'll also need to run electrical wire from the turbine to the house and purchase an inverter (although they're included in most batteryless grid-tie wind turbines). If you're going off-grid or want battery backup for your grid-connected system, you'll also need to buy batteries. All of this will add to the cost. The cost of the turbine itself may range from 10 to 40 percent of the total system cost.

Governing Systems

Found in all wind generators worth buying, governing, or overspeed control, systems are designed to prevent a wind generator from burning out or breaking apart in high winds. They do this by slowing down the rotor when the wind reaches a certain speed, known as the governing wind speed. Why is this necessary?

As wind speed increases, the rotor of a wind turbine spins more rapidly. The increase in the revolutions per minute (rpm) increases electrical output. Although electrical output is a desirable goal, if it exceeds the machine's rated output, the generator could overheat and burn out. In addition, centrifugal forces in high wind speeds exert incredible forces on wind turbines that can tear them apart if the rotor speed is not governed.

A governing system is essential because it allows the turbine to shed extra energy when the winds are really strong. Not all wind turbines come with governing mechanisms, however. Many of the smallest wind turbines, the micro-turbines, with rated outputs of around

Fig. 5.5: *Microturbines. Many microturbines like the Marlec (shown here) have no governing mechanism to slow the rotor in high winds. They rely on the relatively low rotor speed and rugged construction to endure high winds.*

400 watts, for example, have no governing mechanisms (Figure 5.5). (These turbines are too small to produce a significant amount of electricity for most applications.) Larger wind turbines, those with swept areas over 38 square feet, however, come with overspeed controls. Two types are commonly found: furling and blade pitch.

Furling

Most manufacturers protect their wind turbines by furling. Furling is accomplished in one of two ways, both of which shift the position of the rotor (hub and blades) relative to the wind. This turns the blades out of the wind, decreasing the amount of rotor swept area that intercepts the wind. Reducing the swept area reduces the speed at which the rotor turns and the energy collected. Slowing the rotor will protect the wind turbine from damage.

Manufacturers employ two main types of furling: horizontal and vertical. In horizontal furling, the rotor turns out of the wind by turning sideways. For this reason, horizontal furling is also known as side furling. In vertical furling, the rotor rotates upward with the same effect. "Angle furling" is a combination of the two.

Horizontal or side furling is achieved, in part, by hinging the tail. In side-furling turbines, a hinge is located between the tail boom and the body of the turbine. As you can see from Figure 5.6, the turbine is also slightly offset from the yaw axis — that is, the yaw bearing is attached to the side of the turbine body, not its center so the turbine is not directly over the tower. Because the turbine is offset from the yaw axis, the force of the wind on the blades tends

Fig. 5.6: *Side Furling. This wind turbine is not broken, it is side furling in high winds, which slows the rotor and protects the machine from damage.*

MICK SAGRILLO

Fig. 5.7:

Vertical Furling

to rotate the machine around the yaw axis. However, the tail resists this rotation and keeps the rotor facing into the wind.

In light winds, the forces on the rotor and tail are small and the wind holds the tail in its normal position — straight behind the turbine. However, in strong winds, the increasing forces on the rotor overcome the force of the wind on the tail. Since the tail creates more force than the offset rotor, the tail stays mostly aligned with the wind and the turbine turns away from the wind. As a result, the turbine folds on itself. This slows the rotor.

Vertical furling is achieved by moving the hinge in front of the yaw axis and rotating it slightly. In high winds, the force of the wind tilts the rotor up, while the tail stays oriented downwind. As in side furling, this reduces its speed (Figure 5.7).

When fully furled, the rotor of a vertical furling turbine resembles a helicopter rotor. When wind speed declines, however, the rotor returns to its normal operating position. Shock absorbers are often used to ease the rotor back into position.

Furling reduces the amount of energy collected by the rotor. Although electrical output typically continues, it usually occurs at a lower rate, as shown in the power curve of the ARE442 in Figure 5.8.

Changing Blade Pitch

The second type of overspeed control involves a change in the angle of the blades, known as blade pitch, to reduce rotor speed. Blade

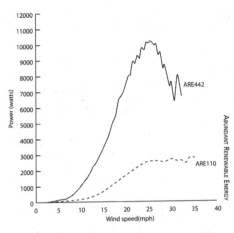

Fig. 5.8:
Power Curves.
The electrical
production of the
ARE442, like that
of many other
turbines, declines
significantly in high
wind speeds as a
result of overspeed
controls that
protect the wind
turbine from

damage. The power production of the ARE110 plateaus, so the machine
continues to produce a significant amount of energy in high winds.

Fig. 5.9: *Pitch Control —*
Blade-Actuated Governor.
Numerous ingenious
methods of blade pitch
control have been
devised. In this turbine, a
Jacobs 31-20, the springs
are part of a complex and
effective blade pitch
control mechanism.

pitch changes automatically in these turbines as wind speed
increases over a certain level. The greater the wind speed above the
operating range of the machine, the more the blades rotate (pitch).
Changing the angle of the blade reduces rotor speed.

Pitch control typically requires springs, gears and weights ingen-
iously arranged to produce the desired effect (Figure 5.9). Some
machines, like the Jacobs, use the weight of the blade itself to
change the pitch.

Blade pitch functions admirably, but is not as widely used as horizontal and vertical furling mechanisms. Of the two, blade pitch control is more expensive, but provides better control of blade speed and is more reliable. Bottom line: although furling mechanisms are cheaper, cheap is not necessarily better when it comes to a wind machine. The goal in buying a wind machine is to purchase the most reliable and most durable turbine. That said, you may only have a few choices among the turbines that produce the amount of electricity you need and most of home-scale wind turbines use furling.

Shut-Down Mechanisms

Small wind turbines should include a reliable shut-down mechanism. They allow a turbine to be turned off so operators can maintain and repair a wind turbine without fear of injury. They also provide a means of shutting a wind machine down when extremely violent storms, especially thunderstorms, are approaching. Maintenance personnel engage the shut-down mechanisms when they need to work on a turbine, but they also typically secure the blades with rope — just in case the wind comes up while they're servicing a turbine.

Wind turbines contain two types of shut-down mechanisms: mechanical and electrical. Mechanical systems include disc brakes and folding tails. Both are manually activated. They're attached to a cable that runs down the tower. Tightening the cable activates the brake or folds the tail (side furling the machine), stopping the rotor (Figure 5.10).

Fig. 5.10:
Cable Winch on Tower. Clay Sterling, MREA's Education Director, shuts down a Jacobs wind turbine by tightening the cable attached to the tail of the turbine.

DAN CHIRAS

Although disc brakes may seem like a good idea, they are not fail-safe. If the cable breaks in violent storm, for example, an operator would be helpless to stop the turbine. There's no way to apply the brakes!

Although folding the tail protects the rotor from overspeeding, it doesn't stop it from rotating. This presents a potential risk to service personnel working on the tower, unless another means of stopping the rotor, such as a disc brake, is available. Furthermore, if the cable breaks in high winds, when the machine is shut down, the tail will swing back into the wind and the wind turbine will start back up. If the winds are strong enough, this could seriously damage the turbine.

Some wind turbines come with electrical brakes, a.k.a. dynamic brakes. Dynamic braking is the least expensive option and is found in many small-scale wind turbines.

Dynamic braking is a fairly simple approach that is found in turbines equipped with permanent magnet alternators. It consists of a switch inside the house or at the base of the tower. When the brake switch is closed, it short-circuits the wind machine, rapidly slowing the rotor.

In dynamic braking, the braking force is proportional to the rotor speed. As the rotor slows down, the braking force diminishes. As the rotor speed approaches zero, so does the braking force. In low to moderate winds, dynamic braking should either stop the rotor or slow it down considerably. However, dynamic braking may not completely stop the rotor in high winds. In winds blowing over 20 miles per hour, for instance, dynamic brakes can't be counted on. If a wind machine is shut down prior to a storm's arrival, strong winds may overpower the brakes, causing the rotors to start turning. In high wind speeds that force the blades to start spinning slowly, energy is dissipated in the windings of the alternator, which could cause it to burn up. Not all dynamic brakes are created equal. Those found in Southwest Windpower's Skystream 3.7 and turbines made by Proven and Abundant Renewable Energy are 100% reliable, as far as we can tell.

Shut-down mechanisms of a wind turbine should be high on the list of considerations, right up there with swept area and tower top weight. If the turbine is to be serviced on the tower, the shut-down mechanism should be capable of completely stopping the rotor. Don't buy a turbine without a shut-down mechanism. Inexpensive wind turbine designs without a reliable shut-down mechanism are a short-sighted gamble, at best.

Sound Levels

The sound a turbine produces is another important factor to consider, both for your own peace of mind and your neighbors'. All residential wind machines produce sound. Sounds emanate from the blades as they spin. They produce a swooshing sound. Sound is also produced when a turbine furls in high winds. Rotation of the rotor in the alternator also produces sound, as do gears in gear-driven wind turbines. (Sound test reports can be found at the National Renewable Energy Laboratory's website.)

Sound levels increase as wind speed increases. However, sound from a wind turbine is often difficult to detect and is rarely a nuisance, as noted in Chapter 1. Remember, too, that mounting a turbine high off the ground — typically 80 to 120 feet — to reach the smoothest, most powerful winds significantly reduces sound levels at ground level.

Even so, it is important to consider sound levels. One way is to observe turbines you are considering in operation under a variety of wind speeds. If you can't, you may want to ask homeowners or business owners who have installed the turbines you are considering for their experiences.

Another method is to check out the rpm of the turbines at their rated outputs. Rated output is the output in watts at a certain wind speed, known as rated speed. Knowing this gives an idea of how much sound they'll produce — the higher the rpm, the more sound. The rpm of a wind turbine also give an indication of quality. Generally, less expensive and less durable turbines spin at a higher rpm. They rely on less expensive generators that operate at

CHAPTER 6

TOWERS AND TOWER INSTALLATION

The tower on which a wind turbine "flies" plays a huge role in a successful wind system. It raises a wind turbine to a nonturbulent height at which it can harvest lots of energy from the wind. A tower also withstands nature's fury.

Tower installation involves a significant investment in money, too. In larger systems, the cost of a tower may account for nearly a quarter of the price tag. For smaller wind turbines, an appropriately sized tower plus installation may cost two to three, sometimes five times, more than the turbine itself.

Tower installations also require a significant amount of time. Most students at wind energy workshops are surprised to learn that most of their time is spent not working on the turbine, but assembling and raising the tower.

In this chapter, we'll explore three tower options, tower assembly and installation. We'll discuss proper siting and the economic benefits of installing a turbine on a tall tower. We'll end with a description of safety concerns and give you some advice on buying a tower.

Tower Options

Towers for small wind machines come in three basic varieties: (1) freestanding, (2) fixed guyed, and (3) tilt-up towers (Figure 6.1). Each type has some variations, listed in Table 6.1.

Table 6.1: Tower Types

1) Freestanding
   a) Lattice
   b) Monopole

2) Fixed Guyed
   a) Lattice
   b) Tubular (typically homebuilt)

3) Tilt-up
   a) Guyed tubular
   b) Guyed lattice

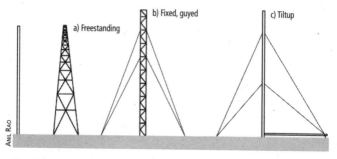

Fig. 6.1: *Wind Tower Options. (a) Freestanding, (b) fixed guyed, and (c) tilt-up. Freestanding towers can be both lattice (shown here) and monopoles. Fixed guyed towers are typically lattice towers. Tilt-up towers can be either lattice or tubular.*

Freestanding Towers

Freestanding wind generator towers are self-supporting structures. They stand on their own, like flag poles or street lights or the Eiffel Tower. Freestanding towers are made of steel and are firmly anchored to the ground via well-reinforced concrete foundations. The combination of heavy-duty steel tower construction and a secure anchorage ensures that the tower can withstand powerful winds that could pry the foundation loose and topple the tower and your expensive turbine. They also ensure that the tower can support the turbine.

The most common type of freestanding tower is a lattice or truss tower, like those shown in Figure 6.2. The Eiffel Tower in Paris is a good example of a lattice structure.

Is Wind Right for You?

Periodic inspection and maintenance and occasional repair of wind turbines are essential to the long-term success of a wind energy system. The towers on which they stand present a formidable barrier to these activities. Many wind system owners fail to perform these tasks because they don't want to lower or climb their towers once a year. If you are a "put it up and forget about it" kind of person and can't afford to hire someone to perform an annual inspection and maintenance, we recommend that you consider installing a PV system instead. PV systems are as close to maintenance-free technology as you can get (provided there are no batteries in the system). If you install a wind system, you will either need to climb the tower or lower it to the ground once or, preferably, twice a year to inspect the turbine, wires, connections, and perform maintenance, as required.

a   b

Fig. 6.2a and 6.2b: *Lattice Tower.* Freestanding lattice towers are made of (a) heavy-duty angle iron, as in this tower erected at the Midwest Renewable Energy Association's headquarters, or (b) tubular steel. Horizontal and vertical bracing made of steel angle iron that runs between the tubular steel legs.

Lattice towers are made from tubular steel or angle iron with horizontal and diagonal cross bracing bolted to the vertical steel legs. Ladders or step bolts are incorporated so the towers can be climbed for inspection, maintenance and repair. Freestanding lattice towers are sometimes fitted with a small platform near the top, which provides a secure place to work.

Another, more expensive option for freestanding towers is the monopole (Figure 6.3). They consist of a single, sturdy pole made from round tubular steel. Rungs or foot pegs are attached for climbing.

Freestanding towers are secured to massive steel-reinforced concrete foundations, as shown in Figure 6.4. The taller the tower, the heftier (and more expensive) the foundation.

Fig. 6.3: *Monopole Tower. Monopole towers are sturdy, well-anchored by a solid foundation, but extremely costly for reasons explained in the text.*

Fig. 6.4: *Concrete Piers and Base of Tower. This massive, deep foundation supports a 120-foot freestanding lattice tower. Each leg of the tower will attach to a steel leg embedded in each of the vertical piers. The piers and base of the foundation are made of concrete reinforced with rebar.*

Assembling and Installing Freestanding Towers

Freestanding lattice towers are typically assembled on the ground in sections, 20 feet at a time. The legs and bracing are bolted together on the ground.

After a lattice tower is assembled, the turbine is often attached. The tower and turbine are lifted with a crane. The tower is bolted to steel anchors embedded in the concrete foundation. To facilitate tower construction, some lattice towers are hinged at the base. That way, the tower can be assembled on the ground, and then tilted up into position with a crane (Figure 6.5). In some instances, the tower is erected and raised without a turbine. The turbine is then hoisted onto the top of the tower. A reasonably level area is needed to assemble a freestanding tower and to lift it with the crane.

Although freestanding lattice towers are typically assembled on the ground and lifted with a crane, it is possible to construct towers vertically one section at a time using a vertical gin pole. This technique is time-consuming and requires extreme caution and is only used in crane-inaccessible sites.

Fig. 6.5: *Crane Lifting Tower and Turbine. This 80-ton crane lifts a massive turbine and tower into place. The hinges at the base of the tower allow the crane to tilt the tower into position.*

Table 6.2
Estimate of Tower Costs for a 120-foot Tower*

Tower Type	Tower Cost	Installation Cost	Total Cost
Freestanding monopole	$25,000	$25,000	$50,000
Freestanding lattice	$15,000	$12,000	$27,000
Guyed tilt-up	$10,000	$4,000	$14,000
Guyed lattice	$8,000	$2,000	$10,000

*Based on 2007 cost data for a typical 10 KW turbine from Mick Sagrillo's Wind Site Assessor Course. These prices include not only the cost of the tower, but also shipping, sales tax, excavation, rebar, concrete, labor to form the foundation, pour the foundation, and strip the forms, and finally backfilling.

Like lattice towers, monopole towers come in sections. They are fitted together on the ground. When completed, the tower is hoisted into place with a crane and the tower is secured to the foundation.

Freestanding monopole towers are typically the most expensive of all options, because they require the most steel and the most robust foundations (Table 6.2).

Pros and Cons of Freestanding Towers

Freestanding towers offer advantages over other types. One of the most important is that they require much less space (Figure 6.6). Their smaller footprint makes a freestanding wind generator tower ideal for locations with extensive tree cover.

Freestanding towers are more aesthetically appealing to many people than guyed towers. A freestanding tower is also one of the safest towers to install. Almost all the work can be done on the ground, and a single crane lift can erect the tower, turbine, wiring, etc.

Embodied energy is the energy that it takes to make a product — from the extraction of the raw materials to the completion of the finished product, including shipping to retail outlets where it is sold. Because they require so much concrete and steel and because these materials require huge amounts of energy to produce, freestanding towers have a much higher embodied energy than other options. If your primary motivation is to decrease your environmental footprint by using renewable energy, a freestanding tower is not your best choice.

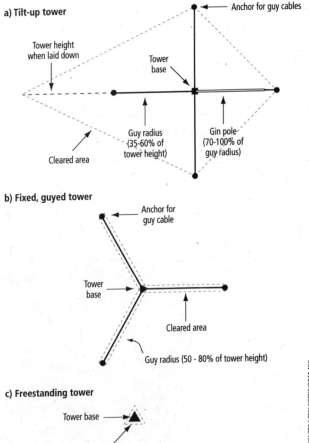

a) Tilt-up tower

Anchor for guy cables

Tower height when laid down

Tower base

Guy radius (35-60% of tower height)

Gin pole (70-100% of guy radius)

Cleared area

b) Fixed, guyed tower

Anchor for guy cable

Tower base

Cleared area

Guy radius (50 - 80% of tower height)

c) Freestanding tower

Tower base

Cleared area

IAN WOOFENDEN AND ANIL RAO

Fig. 6.6: *Tower footprints: (a) Tilt-up, (b) Fixed Guyed, and (c) Freestanding.*

Freestanding towers also require periodic ascent to perform routine inspection, maintenance, and repair, which can be a plus or minus, depending on your point of view. To prevent catastrophic falls, a safety harness or safety work belt must be worn while climbing and working on a tower (Figure 6.7). Safety harnesses are equipped with several D-rings (three-D-ring models should be used for tower work). The D rings are used to secure you to the tower via lanyards to prevent falls when working on a tower. A "positioning" or "restraint" lanyard holds a worker in place to allow him or her to work hands-free. A "shock absorbing" lanyard is used to arrest a fall, that is, gradually slow a worker who has fallen to prevent a harmful jerk.

Towers should be equipped with a safety cable that runs the length of the tower along the climbing rungs or ladder (Figure 6.8). Workers attach their safety harness to the cable when climbing by an anti-fall device, such as a Lad-Saf. This sliding "climbing car" follows you as you ascend but locks onto the cable to arrest a fall if you lose your footing and fall.

Once you are atop the tower, belt in with lanyards and disconnect from the anti-fall cable. You must always be "attached" to the

Fig. 6.7: *Safety Harness. Mick demonstrates proper use of a safety harness in one of his workshops.*

DAN CHIRAS

a   b

Fig. 6.8a and 6.8b: *Safety Cable and Lad-Saf. (a) Worker prepares to climb tower. Note safety harness and Lad-Saf attached to safety cable. This prevents the worker from falling. (b) Close up of connection to Lad-Saf and safety cable.*

tower. When climbing a tower without a safety cable, "Always climb using two lanyards in an alternating pattern so that one of them is clipped onto the tower at all times," advises small wind expert Jim Green.

If you are not willing or able to climb a tower, you must be willing to hire someone to do it. If not, consider installing a tilt-up tower or a PV system.

Fixed Guyed Towers

The second type of tower is the fixed guyed tower (Figure 6.1b). Most are lattice towers. The legs of fixed guyed lattice towers are made of steel tube or pipe, or sometimes solid steel rods. The three legs of the lattice tower are usually 18 inches apart and are secured by horizontal and diagonal steel cross braces (Figure 6.9b).

Guyed towers are bolted to a concrete foundation and are supported by guy cables. Guy cables consist of high-strength stranded-steel cable or aircraft cable. They extend from attachments on the tower to steel-reinforced concrete anchors embedded in the

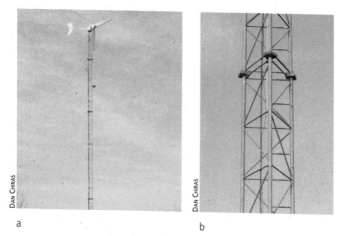

Fig. 6.9a and 6.9b: *Fixed Guyed Lattice Tower. (a) This lattice tower is anchored by guy cables and is one of the most popular and least expensive tower options. (b) Close-up showing details.*

ground. Guy cables are strung out in three directions 120 degrees apart. The guy radius, that is, the distance from the base of the tower to the anchors, ranges from 50 to 80 percent of the tower height, depending on the construction of the tower. Usually it is about 75 percent. For a 100-foot tower, then, the anchors would be 120 degrees apart and 50 to 80 feet from the base.

Fixed guyed towers are also made from pipe or tubular steel that comes in 20-foot sections. Like guyed lattice towers, tubular towers are supported by guy cables.

Assembling and Installing Fixed Guyed Towers

Guyed towers are usually assembled on the ground. Lattice towers are bolted together, one section at a time. After the tower is assembled, the wind turbine and electrical wire are attached. The tower and turbine are then erected by a crane. If the tower is 80 feet or taller, it may be necessary to lift a lattice tower or tower made from steel tubing in sections. The wind turbine is lifted onto the tower after the last section is in place.

Fixed guyed towers can also be assembled vertically, one section at a time, using a vertical gin pole — an inexpensive, temporary vertical "crane" that's bolted onto the tower. Installers use it to raise one section of a tower at a time. After a section is in place, the gin pole is moved up so the next section can be installed, and so on. Vertical gin pole assembly is time-consuming and tedious, and it can be a bit dangerous. Those who've tried it do not recommend it. If no crane is available or the crane cannot access the site, however, a vertical gin pole may be your only option.

Fixed guyed towers rest on concrete pads, though the towers are generally not bolted to them. Guy cables are attached to the tower during assembly. After the tower is upright and plumbed, workers tension the cables. If the turbine was not previously attached, it is then lifted by the crane and fastened to the top.

Pros and Cons of Fixed Guyed Towers

Fixed guyed towers cost much less than freestanding towers because they require much less steel and their foundations require a lot less concrete. Lattice towers used by installers are also mass produced for the telecommunications industry, making them less expensive and widely available. Fixed guyed towers require more space than freestanding towers, but less than tilt-up towers, discussed next.

Fixed guyed towers must also be climbed for routine maintenance and repair, like freestanding towers. Some people consider the guy cables to be an eyesore, although guy wires disappear into the background from most vantage points, except up close. Guy wires may also present a hazard to birds, although we've never heard of a bird killed by them.

Tilt-Up Towers

The third type of tower option is a guyed tilt-up tower. Unlike freestanding and fixed guyed towers, a guyed tilt-up tower can be raised and lowered for inspection, maintenance and repair. Guyed tilt-up towers may be made from steel pipe or lattice sections.

Guyed tilt-up towers require four sets of guy cables at each level. Cables are located 90 degrees apart. The fourth cable is required for stability when raising or lowering a tower. That is, it allows workers to safely raise and lower the tower. Without them, the tower would topple during these operations.

As illustrated in Figure 6.10, a tilt-up tower is raised and lowered with the aid of a gin pole. Unlike the vertical gin pole discussed earlier, this pole is permanently attached to the base of the tower at a 90° angle to the mast. It is a lever arm that allows the tower to be tilted up and down.

Tilting a tower also requires a hinge between the mast and the concrete base (Figure 6.11). When the tower is down — that is, lying on the ground and ready to be raised — the gin pole sticks straight up. When the tower is vertical, the gin pole lies near and parallel to the ground. As illustrated in Figure 6.10, a steel cable connects the free end of the gin pole to a lifting device such as a tractor.

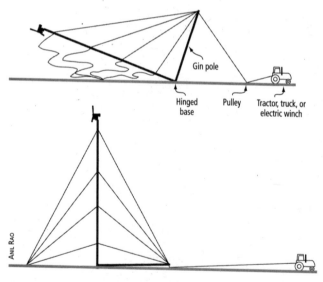

Gin pole

Hinged base Pulley Tractor, truck, or electric winch

ANIL RAO

Fig. 6.10: *Guyed Tilt-Up Tower. Guyed tilt-up towers are raised and lowered using a truck, tractor, electric winch or grip hoist.*

Guy cables hold a tilt-up tower upright and resist the force of the wind. The guy radius is 35 to 80 percent of tower height, depending on the type of tower. For a 100-foot tilt-up tower, the anchors should be located 35 to 80 feet from the base of the tower.

Assembling and Raising a Tilt-Up Guyed Tower

Steel pipe or tubing and lattice towers are both used for guyed tilt-up towers. They come in 20-foot lengths. The individual lengths

Fig. 6.11: *Hinged Base of a Guyed Tilt-Up Tower. The hinge at the base of this tilt-up tower allows it to be tilted up and down to maintain and service the wind turbine.*

Dan Chiras

Dan Chiras

Fig. 6.12: *Gin Pole. The gin pole is attached to the anchor. Notice the electric winch to the left of the attachment.*

of pipe are secured by bolts or joined by slip-fit couplings on the ground. While the tower is on the ground, guy cables are attached to the tower and the concrete anchors.

Once assembled, the tower is tilted into position (Figure 6.10). This is accomplished with the assistance of a tractor, a pickup truck, a heavy-duty electric winch or a manually operated device known as a grip hoist.

When installing a tall tower for the first time, some installers raise one or two sections of the tower at a time. After each section is raised, the tower is plumbed and the guy cables are tensioned. This tower is then lowered and an additional piece is added. It is then raised, plumbed, and cables tensioned. This continues until the entire tower is assembled, plumbed and properly tensioned.

Experienced installers also recommend lifting (and plumbing) the entire tower *before* attaching the wind turbine to be sure that everything is correct. Also, make sure to train workers so they all know what they are doing by the time the turbine and tower are lifted.

Pros and Cons of a Guyed Tilt-Up Tower

The main benefit of tilt-up towers is that they never have to be climbed. They can be raised and lowered fairly quickly and all inspections and work can be performed on terra firma.

Although they're ideal for those who cringe at the idea of climbing a tall tower, tilt-up guyed towers have the largest footprint of all (Figure 6.6). You'll need to ensure that there's a clear path for the lifting vehicle and a lay-down zone as long as the tower.

Raising and lowering a tower requires a few helpers to ensure that everything runs smoothly — for example, that the cables don't get tangled. And, of course, you'll need a truck, tractor or some other lifting device. Be careful when using a tow vehicle because they can slip. Accidents can also occur if the anchors are not correctly positioned or the guy cables get too tight while lowering or raising the tower. A strong wind could come along and blow the tower over when it is being raised or lowered, ruining the turbine.

Tower Kits

Several turbine manufacturers sell tilt-up tower kits designed and engineered for their wind machines. Because steel pipe is heavy and expensive to ship long distances, most kits include all of the materials you need *except* the pipe. You purchase pipe locally.

Wind turbines often require adaptors, known as stub towers, to fit onto commercially available towers. Stub towers consist of a short piece of pipe, with a flange that bolts onto the base of the turbine. Be sure to obtain an adaptor for your turbine. For best results, buy a hot-dipped galvanized adaptor. They last much longer than ungalvanized steel sub-towers.

When buying a tower, be sure you purchase a model that's approved by the wind turbine manufacturer. When shopping for a wind generator tower kit, also be aware that some kits may not include anchors, because the type of anchor needed for a tower varies from one location to the next, depending on the soil type. In most cases, anchor foundations are constructed on site using concrete and rebar.

When mounting a wind turbine, be sure to avoid roof-top towers. Such installations are an unequivocally bad idea for several reasons. First, roof-mounted turbines in most locations are too close to the ground, where average annual wind speeds are much lower. Lower average annual wind speeds mean much lower output. A 2007 study of roof-mounted wind turbines by Encraft Ltd, a British consulting firm, showed that wind turbines mounted on homes and apartment buildings produced much less electricity than predicted by computer models. On residential structures, annual wind speeds were only about 5 to 7 miles per hour, well below the models' predicted wind speeds of 10 to 12 miles per hour — and well below the speed at which the wind turbines produce electricity. (The models obviously didn't account for ground drag and turbulence).

Meters were installed to monitor the output of the micro-turbines and small turbines in this study. The researchers compared energy generated by the urban turbines to the manufacturers' performance predictions. The researchers found that on average, wind turbines exported less than 0.5 kilowatt-hours a day — that's 5 cents worth of electricity per day. Some of the turbines generated less power than the inverters consumed and were negative energy producers at their low wind speed sites. That is, these systems consumed more electricity than they generated. Overall, the turbines produced only 50 to 60 kWh of electricity per year — that's $5 to $6 per year!

Rooftop installations are not a good idea because they're also exposed to turbulence created by ground clutter, trees and buildings. As noted in previous chapters, turbulence reduces the quality and quantity of wind. This reduces energy production and increases wear and tear on the turbine, resulting in shorter turbine life.

Buildings are also rarely designed and engineered to support the load and handle the vibrations produced by a wind turbine. These vibrations can cause structural damage to buildings and vibrations are conducted into the building, which can be annoying to occupants.

Tower Anchors and Bases

When installing a tower, you'll need to install a strong base. If the tower is guyed, you'll also need to install anchors for the guy cables. All major manufacturers provide well-engineered plans for suitable foundations. Follow the plans very carefully. Don't cut corners to save time or money. Take a wind workshop or two *before* installing your wind turbine or hire a professional.

Tower Base

Virtually all towers are mounted on concrete pads reinforced with rebar (Figure 6.13). The depth required for a foundation, known as the critical depth, is the depth that prevents a foundation (and anchors) from being pried out of the ground by the force of the wind or uprooted by freeze-thaw cycles. The critical depth of a foundation depends on many factors, such as the type and height of the tower, wind speed, depth of the frost line and soil characteristics.

Fig. 6.13: *Concrete Base. Most wind turbine towers rest on a solid concrete base. This photo shows the base for each leg of a free-standing lattice tower.*

Freestanding towers require deeper and more robust foundations than guyed towers. The taller the tower, the stronger the foundation. The stronger the winds, the deeper and more robust the foundation. The deeper the frost line, the deeper the foundation. Some soils hold tower foundations in place better than others. For instance, clay-rich and heavier soils have more holding power than sandy soils. For advice on critical tower depth, contact the turbine and tower manufacturer.

Engineered plans are typically provided in a turbine's installation manual. Even so, be sure to consult local building codes, soil engineers or local excavators to be sure your tower foundation is sufficiently strong.

For maximum strength, concrete bases should cure at least 28 days prior to installation of the tower. Do not place a tower on a concrete pad before that! Inadequately cured concrete weakens the strength of the concrete. Your tower and wind machine rely on a sturdy, well-cured foundation. Remember, too, that cold weather slows the curing process. More time may be required in such conditions.

Anchors

In addition to a concrete base, fixed guyed towers and guyed tilt-up towers also require anchors to attach the guy cables. Anchor options for wind generator towers are many and varied. For most wind turbines, concrete anchors are the best choice. They are created by digging a deep hole, installing rebar according to the engineered drawing, and then pouring concrete into the hole. A screw-in auger, anchor, or angle steel is then embedded into the hole and angled towards the tower as per manufacturer instructions. Once the concrete has cured, the hole is filled in. Be sure to pour anchors well below the frost line.

Proper Siting of a Wind Machine

A wind turbine must be mounted in a good wind site, well above ground clutter in the strongest, smoothest winds. Wind site assessors begin the process of siting a wind turbine by determining the prevailing wind direction at a site. Although winds blow in different directions at different times of the year, or even within the same day, they arrive from one or two directions predominantly over the course of the year. In many places in North America, winds come predominantly from the southwest — thanks to the Coriolis effect (Chapter 2). They often blow from the northwest in the winter.

To determine the predominant wind flow, ask the advice of farmers, who work outdoors and hence are familiar with wind patterns, or contact a local airport. They may be able to provide you with a wind rose, a graphical representation of wind direction (Figure 6.14). In a wind rose, the length of the spokes around the circle is an indication of how frequently the wind blows from a particular direction. The longer the line, the greater the frequency. In the wind rose in Figure 6.14, the winds blow predominantly from the southwest. A wind rose also indicates the percentage of total wind energy from each direction, which is very helpful. You can also find data on wind direction at the NASA Surface Meteorology site discussed in Chapter 4.

In an open site, with little ground clutter, a wind turbine can be located almost anywhere — so long as the entire rotor is mounted 30 feet above the tallest obstacle within a 500-foot radius and you've taken into account future tree growth, if trees are the tallest objects.

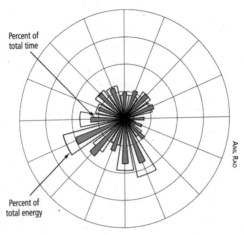

Fig. 6.14: *Wind Rose. This unique graph shows how often winds blow from various directions and the percent energy of the wind for various directions. The wider white bars represent the percent of total energy from different directions and the narrower, shaded bars illustrate the percent of total time from each of the sixteen different direction sectors.*

Unfortunately, very few of us live on ideal sites. There's almost always some major obstacles.

To site a wind turbine, first determine the prevailing wind direction, then look for a location for the tower that's upwind of major obstacles. Although winds will shift so that upwind temporarily becomes downwind, situating your wind turbine and tower this way will ensure that it can take advantage of the strongest prevailing winds.

When siting a wind machine, it is also a good idea to minimize wire runs from the turbine to the controller and inverter to reduce line loss. As a rule, the higher the wind turbine's voltage, the farther it can be sited from the point of use. When installing a turbine, contact the manufacturer or an experienced installer for recommendations.

Tower Height Considerations

Once you've identified the best site, you must determine optimum tower height. To produce as much electrical energy as possible at a site, the rule of thumb is that the *entire* rotor should be *at least* 30 feet above the tallest obstacle within a radius of 500 feet.

When calculating minimum tower height, don't forget to take tree growth into account. If the trees on your property will grow 20 feet in the next 20 to 30 years, the life expectancy of a wind system, add that to the tower height for the best long-term performance. (To learn two ways to estimate the height of trees and buildings, see the accompanying box.)

How High?

Determining the height of a tower seems pretty straight-forward until you have to do it. The first challenge you'll face is determining the actual height of nearby objects, such as trees or barns. How do wind site assessors determine the height of ground clutter?

One way, shown in Figure 6.15, is to place a stake (a metal fence post, for instance) next to the object you want to measure. On a sunny day, measure the height of the stake and then measure its shadow. Then measure the shadow of the object under question.

The height of the object can be determined by ratios using the equation: $\frac{H_1}{H_2} = \frac{SL_1}{SL_2}$. H_1 is the unknown height and H_2 is the height of the fence post. SL_1 is the length of the shadow of the object you are trying to measure. SL_2 is the length of the shadow of the fence post. To solve for H_1, you just need to rearrange the equation: $H_1 = \frac{(SL_1)(H_2)}{SL_2}$. Note that the ground around both the tree and the fence post must be level for this method to be accurate.

Consider an example. Let's assume that the fence post is four feet high and the shadow it casts is two feet long. The shadow cast by the tree or building is 14 feet. How high is the tree? As illustrated, in Figure 6.15a, you begin by setting up ratios: $\frac{x}{4} = \frac{14}{2}$, then solve for x: $x = (4 \times \frac{14}{2}) = 28$ feet. Another simple method is explained in Figure 6.15b. ☛

If your site is within a quarter of a mile from a forest or good-sized wooded lot, the top of the nearby tree line is the height you want to exceed. Mount the wind turbine using the tree line as the height you must exceed. Don't forget to factor in tree growth.

If you are installing a wind turbine in an area with more than 50 percent deciduous tree cover, the effective ground level is two thirds of the tree height. If trees are 60-feet high, for instance, the

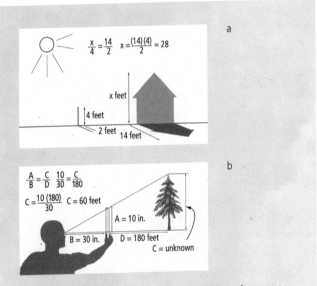

Fig. 6.15a and 6.15b: *Measuring Height. (a) Driving a fence post or some other object of known length into the ground next to an object of unknown height and comparing the length of the shadows allows one to calculate the height of an object. (b) Another method for determining height is shown here. In this method, you'll be solving for C, the height of the tree. Measure the distance from the tree (D). Measure the distance from your eye to the ruler in your hand. This is B. Measure the height of the object in inches on the ruler. This is A. Then set up a ratio equivalence as follows:*
$\frac{A}{B} = \frac{C}{B+D}$. *The rest of the math is shown in the figure.* ∎

effective ground level is 40 feet. A 100-foot equivalent tower would, therefore, need to be 140 feet high to take into account the trees.

Bear in mind that the height recommendation is the *minimum* acceptable tower height. Savvy wind energy installers exceed the rule and see increased performance because of it. It usually costs very little to increase tower height by another 20 to 40 feet and the return on this small investment is quite impressive. We don't know anyone who has installed a wind turbine who says, "I wish I'd bought a shorter tower." However, we know lots of people who wish they had purchased a taller one.

Tall Tower Economics:
Overcoming the Small-Turbines-on-Short-Tower Myth

When you talk to professional wind system installers, you may hear statements to the effect that it doesn't make sense to mount a smaller turbine, for example, one with a seven- or eight-foot diameter rotor, on a tall tower. This is flawed reasoning. Tower height should be determined by the height of obstructions in the area, not the size of the wind turbine or the towers a manufacturer or dealer sells. A 50-foot tower slightly downwind from a 65-foot-high tree line isn't going to produce much electricity. Moreover, the turbine will produce even less electricity as the trees grow over the 20- to 30-year life of the wind system. Remember: energy output and the economics of the wind system are both proportional to V^3 (the cube of the wind speed).

Although it is sometimes hard to justify a tall tower for a small turbine, that doesn't mean that the right decision is a short tower. The right decision is to invest enough in your tower to make the most of your turbine's potential — or choose another renewable energy system.

If you are thinking about installing a smaller wind generator, but are nervous about the cost of a taller tower, we recommend that you calculate how much more the tower will cost and how much more electricity the turbine will produce on a taller tower. In our experience, installing a taller tower always results in the production of substantially more electricity. Even though it will always cost more money, the important question to ask is whether the increased tower height is justified economically by the increase in electrical production. In most cases, it is.

Aircraft Safety and the FAA

Another factor to consider when installing a wind turbine is its impact on aviation. If the tower exceeds a certain height and is within a certain distance from an airport, you will need to file for a permit from the Federal Aviation Administration (FAA). As for height, an FAA permit is required if the tower is over 200 feet, which is extremely rare for a small wind turbine.

The second condition that requires an FAA review and permit is if the wind generator is approximately two to four miles from a "public use" or military airport. Whether you need a permit depends on the length of the runway. In such instances, the FAA will determine the height of the tower you can install. They may also require top-of-tower warning lights. Note that permits are not required when siting a turbine near private landing strips with no public access, airfields not shown on FAA maps, or landing strips that are not in use. If you're hiring a professional installer, he or she can advise you on this matter.

Protecting Against Lightning

Although lightning is not attracted to tall metal objects, such as a wind generator tower, as is commonly thought, it is important to install lightning protection. Grounding rods attached to the tower bleed off the static charge created as air masses move across the Earth's surface. They'll reduce the likelihood of a direct strike. Ground rods are eight-foot long copper-coated rods. They are driven into the ground at the base of the wind turbine tower. For details, contact a local installer or your turbine/tower manufacturer. (For more on the subject, see Mick Sagrillo's article "Residential Wind Turbines and Lightning," available online at renewwisconsin.org/wind/Toolbox-Fact%20Sheets/Lightning.pdf.)

Grounding a tower minimizes lightning strikes, but does not guarantee that lightning will not strike your tower. Backup is needed in the form of lightning arrestors. Lightning arrestors "bleed off" electricity in case of a direct or nearby strike, protecting sensitive equipment. A professional installer will provide recommendations.

Surge arrestors should also be installed on the electrical wire running down the tower and the utility wiring for grid-tied systems. They protect against surges of electricity induced in the wire by lightning strikes. The surge protectors on the utility side of the system protect against lightning strikes in utility lines, which are much more frequent than on properly grounded wind turbine towers.

While protecting against direct lightning strikes is important, nearby strikes pose the gravest danger to a wind system. When lightning strikes the ground near a home, it creates an electrical current in the atmosphere and/or the ground. This current creates a voltage wave that resembles ripples in a quiet pond after a pebble is dropped into it. If one of these waves crosses a conductor like a wind generator tower or buried wires, an electrical current will be created (induced) in the conductor. This current can fry sensitive electronic components of a wind system. Surge protectors in a wind system will help protect against this phenomenon.

CHAPTER 7

BATTERIES AND CHARGE CONTROLLERS

If you are installing an off-grid system or a grid-connected system with battery backup, you'll need a battery bank and you'll need to know how to install it properly and how to maintain it. In this chapter, we'll help you understand your options, how batteries work, and how they should be installed and maintained.

Which Types of Batteries Work Best?

Batteries used in most off-grid renewable energy systems are deep-cycle, flooded lead-acid batteries. These batteries can be charged and discharged (cycled) hundreds of times before they wear out.

Lead-acid batteries contain three separate 2-volt compartments, known as cells. Inside each cell is a series of thick, parallel lead plates (Figure 7.1). The cells are connected internally (wired in series) so that they produce 6-volt electricity. The cells are filled with sulfuric acid (hence the term "flooded"). A partition wall separates each cell so that fluid cannot flow from one cell to the next. The cells are encased in a heavy-duty plastic case.

As illustrated in Figure 7.1, lead-acid batteries contain two types of plates: positive and negative. The positive plates connect to a positive metal post or terminal; the negative plates connect to a negative post. The posts allow electricity to flow into and out of batteries.

The positive plates of lead-acid batteries are made from lead dioxide (PbO_2). The negative plates are made from pure lead. The

Fig. 7.1: *Anatomy of a Flooded Lead-Acid Battery.*

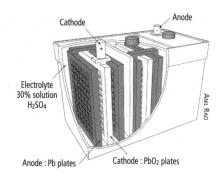

Cathode

Anode

Electrolyte
30% solution
H₂SO₄

ANIL RAO

Anode : Pb plates

Cathode : PbO₂ plates

sulfuric acid that fills the spaces between the plates is referred to as the electrolyte.

How Lead-Acid Batteries Work

Like all other types of batteries, lead-acid batteries convert electrical energy into chemical energy when they are charged. When discharging, that is, giving off electricity, chemical energy is converted back into electricity. Electricity, of course, consists of electrons, tiny negatively charged particles, that flow through conductors. The electrons flow out of the battery at the positive post, creating an electrical current. During the chemical reactions that take place during discharge, lead on the surface of the negative plates reacts with sulfuric acid in the battery, creating tiny lead sulfate crystals on the surface of the plates.

Although the chemistry of lead-acid batteries is a bit complicated, it is important to remember that this system works because electrons can be stored in the chemicals within the battery when a battery is charged. The stored electrons can be drawn out by reversing the chemical reactions. Through this reversible chemical reaction, the battery is acting as a "charge pump," moving electrical charges through a circuit on demand.

Will Any Lead-Acid Battery Work?

Lead-acid batteries come in many varieties, each one designed for a specific application. Car batteries, for example, are designed and manufactured for use in cars, light trucks and vans; deep-cycle marine batteries are designed for boats; golf cart batteries are for golf carts; and forklift batteries for forklifts.

For off-grid systems, you have three options: (1) deep-cycle flooded lead-acid battery like those made by Trojan, Rolls and Deka (Figure 7.2), (2) forklift batteries, and (3) golf cart batteries. Car batteries won't work. Their thin lead plates are not designed for the deep discharges that commonly occur in renewable energy systems. Although the lead sulfate crystals that form on the plates of a battery during deep discharge are removed when batteries are recharged, some crystals fall off before recharge occurs. The thin plates of a car battery are whittled away to nothing very quickly. After twenty or so deep discharges, the batteries would be ruined — no longer able to accept a charge.

Battery-based wind and solar systems require deep-discharge lead-acid batteries with thick lead plates. It's the thickness of the plates that allows them to withstand multiple deep discharges. Even though the plates lose a little lead over time, they are so thick that the small losses are insignificant. Consequently, deep-cycle batteries can be deeply discharged hundreds, sometimes a few thousand times, over their lifetime.

Fig. 7.2: *Deep-Cycle Lead-Acid Batteries. These batteries contain thick lead plates and are used in many battery-based renewable energy systems. The thick plates permit deep cycling so long as the batteries are recharged soon after each deep discharge.*

SURRETTE BATTERY COMPANY LTD.

For optimum long-term performance, deep-cycle batteries still need to be recharged promptly after deep discharging. Don't forget this! With proper care, these batteries could last for seven to ten years, perhaps longer.

Forklift batteries are high-capacity, deep-discharge batteries designed for a fairly long life and operate under fairly demanding conditions. They can withstand 1,000 to 2,000 deep discharges — more than many other deep-cycle batteries used in renewable energy systems — and thus work well. They are, however, rather heavy, bulky and expensive. If you can acquire them new at a decent price, you may want to use them.

Golf cart batteries may also work. Like forklift batteries, golf cart batteries are designed for deep discharge. However, they typically cost a lot less than other heavier duty deep-cycle batteries. While the lower cost may be appealing, golf cart batteries don't store as much electricity and don't last as long as the alternatives. They may last only five to seven years, if well cared for. Shorter lifespan means more frequent replacement. More frequent replacement means higher long-term costs and more hassle.

What about Used Batteries?

Another option is used batteries. Although they can often be purchased inexpensively, they're rarely worth it. Used batteries are often being sold because they've failed or have experienced a serious decline in function. As a buyer, you also have no idea how well — or how poorly — they've been treated. Have they been deeply discharged many times? Have they been left in a state of deep discharge for long periods? Have they been filled with tap water rather than distilled water? Although there are exceptions, most people we know who've purchased used flooded lead-acid batteries have been disappointed.

When shopping for batteries for a renewable energy system, look for high-quality deep-cycle batteries. Although you might be able to save some money by purchasing cheaper alternatives, including used batteries, frequent replacement is time consuming.

Batteries are heavy and it takes quite a lot of time and effort to disconnect old batteries and rewire new ones. Bottom line: the longer a battery will last — because it's the right battery for the job and it's well made and well cared for — the better!

Sealed Batteries

Grid-connected systems with battery backup often incorporate another type of lead-acid battery, known as sealed lead-acid batteries or captive-electrolyte batteries. Sealed batteries are filled with electrolyte at the factory, charged, and then permanently sealed. This makes them easy to handle. They can be shipped without fear of leaking. They won't leak even if the battery casing is cracked, and they can be installed in any orientation — even on their sides. But most important, they never need to be watered.

Two types of sealed batteries are available: absorbed glass mat (AGM) batteries and gel cell batteries. In absorbed glass mat batteries, thin absorbent fiberglass mats are placed between the lead plates. The mat consists of a network of tiny pores that immobilize the battery acid. These tiny pockets also capture hydrogen and oxygen gases given off by the battery when it is charging. Unlike a flooded lead-acid battery, the gases can't escape. Instead, they recombine in the pockets, reforming water. That's why sealed AGM batteries never need watering.

In gel batteries, the sulfuric acid electrolyte is converted to a substance much like hardened Jell-O by the addition of a small amount of silica gel. The gel-like substance fills the spaces between the lead plates.

Sealed batteries are also known as "maintenance-free" batteries because fluid levels never need to be checked and because the batteries never need to be filled with water. They also never need to be (and should not be!) equalized, a process discussed shortly. Eliminating routine maintenance saves a lot of time and energy. It makes sealed batteries a good choice for grid-connected systems with battery backup. In these systems, batteries are rarely used and maintained like those in a grid-connected system. Sealed batteries are also ideal off-grid systems in remote locations where routine maintenance is problematic — for example, rarely occupied backwoods cabins.

Sealed batteries offer several additional advantages over flooded lead-acid batteries. They charge faster and do not release explosive gases, so there's no need to vent battery rooms or battery boxes where they're stored. In addition, sealed batteries are much more tolerant of low temperatures. They can even handle occasional freezing, although this is never recommended. Sealed batteries self-discharge more slowly than flooded lead-acid batteries when not in use. (All batteries self-discharge when not in use.)

Unfortunately, sealed batteries are much more expensive, store less electricity, and have a shorter lifespan than flooded lead-acid batteries. They also can't be rejuvenated (equalized) if left in a state of deep discharge for an extended period. During such times, lead sulfate crystals on the plates begin to grow. Large crystals reduce a battery's ability to store electricity. Batteries then take progressively less charge and have less to give back. Over time, entire cells may die, substantially reducing a battery's storage capacity.

Large crystals on the plates of flooded lead-acid batteries can be removed by a controlled overcharge, a procedure known as *equalization*. Although equalization is safe in unsealed flooded lead-acid batteries, it results in pressure buildup inside a sealed battery. Pressure is vented through the pressure release valve on the sealed battery, which releases electrolyte and could destroy or seriously decrease the storage capacity of the sealed battery. So, while maintenance-free batteries may seem like a good idea, they are not suitable for many applications.

Wiring a Battery Bank

Batteries are wired by installers to produce a specific voltage and amp-hour storage capacity. Small renewable energy systems — for example, those used to power RVs, boats and cabins — are typically wired to produce 12-volt electricity. The electronics in these applications run entirely off 12-volt DC electricity. Systems in off-grid homes and businesses are typically wired to produce 24- or 48-volt DC electricity. The low-voltage DC electricity, however, is converted to AC electricity by the inverter. It also boosts the voltage to 120- and 240-volts, commonly used in homes and businesses.

Sizing a Battery Bank

Properly sizing a battery bank is key to designing a reliable off-grid system. The principal goal when sizing a battery bank is to install a sufficient number of batteries to carry your household or business through periods when the wind or wind and sun (in hybrid systems) are not available.

Battery banks are typically sized to meet the need for electricity for three days. Longer reserve periods — five days or more — may be required for some areas. As noted in Chapter 3, backup fossil-fuel generators are often included in off-grid systems. Backup generators can reduce the size of the battery bank and are used to equalize the batteries. For more details on wiring and sizing battery banks for off-grid systems, you may want to check out Dan's book, *Power from the Wind*. Because batteries are expensive, it's a good idea to make your home as efficient as possible. This will reduce the size of your wind system and battery bank.

Battery Maintenance and Safety

Battery care and maintenance are vital to the long-term success of battery-based renewable energy systems. Proper maintenance increases the service life of a battery. Because batteries are expensive, longer service life results in lower operating costs over the long haul. The longer your batteries last, the cheaper your electricity will be.

Keep Them Warm

Lead-acid batteries like to be kept warm. For optimal function, batteries should be kept between 75 to 80° F. In this range, they'll accept and deliver tons more electricity. Guaranteed! Cold temperatures slow down the chemical reactions in batteries, reducing the amount of electricity a battery can store.

Although batteries can't be housed in cold rooms, care must be taken to avoid exposure to high temperatures as well. High temperatures increase the release of explosive hydrogen gas, known as *outgassing*. They also increase water loss, which reduces battery fluid levels. Higher temperatures also lead to higher rates of self-discharge in batteries. (As a rule, older batteries lose charge faster than new batteries.)

If you can't maintain batteries in a 75 to 80° F range, at least ensure they're housed in a room where the temperature ranges between 50 and 80° F. Rarely should batteries fall below 40° F or exceed 100° F. Whatever you do, don't store batteries in a cold garage, barn or shed. Besides delivering less electricity, they won't last long. They could even freeze under certain conditions, causing their cases to crack, spilling acid and creating a dangerous mess.

Batteries should not be stored on concrete floors. Cold floors cool them down and reduce their rate of chemical reaction and their capacity. Always raise batteries off the floor.

Ideally, batteries should be housed in a separate, conditioned (heated and cooled) battery room or in a battery box inside a conditioned space to maintain proper temperature. Battery boxes are typically built from plywood. An acid-resistant liner is required to contain possible acid spills. Lids should be hinged and sloped to discourage people from storing items on them. As a side note, batteries should be located as close to the inverter and other power conditioning equipment as possible. Doing so minimizes power losses.

Ventilate Your Batteries

Batteries release potentially explosive hydrogen gas when being charged, so battery boxes and battery rooms containing flooded lead-acid batteries should be well ventilated (Figure 7.3). This allows hydrogen to escape. Never place batteries in a room with a gas-burning appliance or an electrical device, such as a water heater, even if the enclosure is vented. A tiny spark could ignite the hydrogen gas, causing an explosion. Note that this applies to the inverter

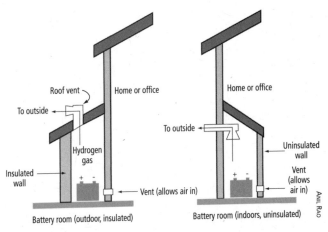

Fig. 7.3: *Battery Vent System. (a) An outdoor battery room should be well insulated and possibly heated and cooled to maintain temperatures in the optimum range. A passive vent system is needed to allow hydrogen gas to escape. (b) Indoor battery rooms need not be insulated, but require venting.*

and all other conditioning equipment. Although you want them *near* the batteries, they should *not* be in the *same* space.

Keep Kids Out

Battery rooms and battery boxes should be inaccessible and locked if young children are present. This will prevent children from coming in contact with the batteries, risking electrical or acid burns. Although electrocution is not a hazard at 12, 24 or 48 volts, dropping a tool or other metal object on the battery terminals could result in an electrical arc that can cause burns, or could result in an explosion of the battery, resulting in acid burns.

Avoiding Deep Discharge to Ensure Longer Battery Life

Keeping flooded lead-acid batteries warm and topped off with distilled water to replace water lost during charging ensures a long life span. Longevity can also be ensured by keeping batteries as fully charged as possible. Like many technologies, lead-acid batteries last longer the less you use them. That is to say, the fewer times a battery is deeply discharged, the longer it will last.

This topic (like so many others) is complicated. While deep discharging reduces the lifespan of a battery, what renewable energy users want from batteries is not simply for them to last a long time but to cycle a lot of energy. Theoretically, you'll get the most bang for your buck by cycling in the 40 to 60 percent deep-discharge range.

It's also important to recharge batteries as quickly as possible after deep cycling. For long life, you should also never leave batteries at a low state of charge for a long time. This results in the formation of large lead sulfate crystals, described earlier. Unfortunately, achieving these goals is easier said than done. If your system is small and you don't pay much attention to electrical use, you'll very likely overshoot the 40 to 60 percent mark time and time again.

One way of reducing deep discharge is to conserve and use electricity efficiently. Conserving energy means not leaving lights and electronic devices running when they're not in use. It also means getting rid of phantom loads. Energy efficiency means installing energy-efficient lighting, appliances, electronics, etc. You can also adjust electrical use according to the state of charge of your batteries — in other words, cut back on electrical usage when batteries are more deeply discharged and shift demand for electricity to times when the batteries are more fully charged. You may, for instance, run your washing machine and microwave when the wind's blowing and your batteries are full, but hold off when batteries are running low.

To track battery state of charge you can install a digital amp-hour or watt-hour meter. These meters keep track of the amount of electricity stored in a battery bank each day. They also indicate the amount of electricity drawn from the batteries. In addition, they keep track of the total amount stored in a battery bank at any one time — how full the batteries are. This information is used to adjust consumption. If batteries are approaching the 40 to 60 percent discharge mark, you may hold off on activities that consume lots of electricity. Or, you may run your backup generator to charge the batteries.

Watering and Cleaning Batteries

To maintain batteries you must also periodically add distilled water. This replaces water lost during charging. Water loss occurs by electrolysis, the splitting of water molecules in the electrolyte when electricity flows into a battery. Electricity splits water molecules into hydrogen and oxygen. (Electrolysis is the source of the potentially explosive mixture of hydrogen and oxygen gas that makes battery room venting necessary.)

Hydrogen and oxygen produced during electrolysis are both gases. These gases can escape through the vents in the battery caps in flooded lead-acid batteries, lowering water levels. Water can also evaporate through the vents in a flooded lead-acid battery at any time, and a mist of sulfuric acid can escape through the vents during charging, depleting fluid levels.

All of these sources of water loss add up over time and can run a battery dry. When the plates are exposed to air, they quickly begin to corrode. When this happens, a battery's life is pretty well over.

To prevent batteries from running dry, check battery fluid levels regularly. Many experts recommend checking batteries monthly. Others recommend checking batteries every two to three months. When replacing fluid, be sure to only add distilled or deionized water, and do not *overfill* batteries. Never use tap water. It may contain minerals or chemicals that will contaminate the battery fluid, reducing a battery's life span.

The tops and terminals of batteries may also need to be cleaned with distilled water and paper towels or a clean rag. When cleaning batteries, be sure to wear gloves, protective eyewear and a long-sleeved shirt you don't care about. If you get acid on your skin, wash it off immediately with soap and water.

When filling batteries, *be sure to take off watches, rings and other jewelry*, especially loose-fitting jewelry. Metal jewelry will conduct electricity if it contacts both terminals of a battery. Such an event will leave your jewelry in a puddle of metal — along with some of your flesh. One 6- or 12-volt cell can produce more than 8,000 amps if the positive and negative terminals of a battery are connected. In addition, sparks could ignite hydrogen and oxygen gas in the vicinity, causing an explosion. Shorting out a battery can also crack the case, releasing battery acid.

Also be careful with tools when working on batteries — for example, tightening cable connections. A metal tool that makes a connection between oppositely charged terminals on a battery may be instantaneously welded in place. The tool will become red hot and could also ignite hydrogen gas, causing an explosion. Wrap hand tools used for battery maintenance in electrical tape so that only one inch of metal is exposed on the working end; that way it can't make an electrical connection. Or buy insulated tools to prevent this from happening. Try to have a set of these insulated tools dedicated just to battery maintenance. That way you won't grab the insulated wrench for another project and then use whatever non-insulated wrench is at hand when it comes time for battery maintenance.

You may also have to clean the battery posts every year or two. To clean the posts, use a small wire brush, perhaps in conjunction with a spray-on battery cleaner purchased at a hardware store. To reduce maintenance, coat battery posts with Vaseline or a battery protector/sealer, available at hardware and auto supply stores. This protects the posts and the nuts that secure the battery cables on the posts.

Equalization

To get the most out of batteries, you need to periodically equalize them. Equalization is a controlled overcharge of batteries.

Why Equalize?

Periodic equalization is performed for three reasons. The first is to drive lead sulfate crystals off the lead plates, preventing the formation of larger crystals that reduce battery capacity. In addition to coating the plates, the large crystals can also flake off, removing lots of lead from the plates.

Batteries must also be periodically equalized to stir the electrolyte. Sulfuric acid tends to settle near the bottom of the cells in flooded lead-acid batteries. During equalization, hydrogen and oxygen gases released by the breakdown of water (electrolysis) create bubbles. They mix the fluid so that the concentration of acid is equalized throughout each cell of each battery, ensuring better function.

Equalization also helps bring all of the cells in a battery bank to the same voltage. That's important because some cells sulfate more than others. As a result, their voltage may be lower. A single low-voltage cell in one battery reduces the voltage of the entire string. In many ways, then, a battery bank is like a camel train. It travels at the speed of the slowest camel.

Although equalization removes lead sulfate from plates, which restores function, some lead flakes off the plates during equalization and settles to the bottom of the batteries. As a result, even properly equalized batteries lose lead over time and never regain their full capacity.

How to Equalize

Equalizing batteries is a simple process. In those systems with a gen-set for backup, the owner simply sets the inverter to the equalization mode and then cranks up the generator. The inverter controls the process from that point onward. In wind/PV hybrid systems, the operator can also set the controller to the equalize setting during a storm or period of high wind. The controller takes over from there.

How often batteries should be equalized depends on whom you talk to and how hard you work your batteries. Some installers recommend equalization every three months. If your batteries are frequently deep discharged, however, you may want to equalize more frequently. If batteries are rarely deep discharged, they'll need less frequent equalization. For example, batteries that are rarely discharged below 50 percent may only need to be equalized every six months.

Rather than second guess your batteries' needs for equalization, it is wise to check the voltage of each battery, using a digital volt meter (multimeter), every month or two. If you notice that the

Fig. 7.4: *Hydrometer. Hydrometers measure specific gravity. Low specific gravity indicates that the battery needs recharging, perhaps even equalization.*

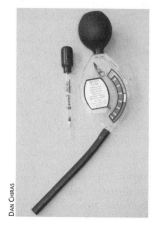

voltage of one or two batteries is substantially lower than others, it's time to equalize.

Another way to test batteries is to measure the specific gravity of the battery acid using a hydrometer (Figure 7.4). Specific gravity is a measure of the density of battery acid. Density is related to the concentration of battery acid — the higher the concentration, the higher the specific gravity. If significant differences in the specific gravity of the battery acid are detected in the cells of a battery bank, it is time to equalize.

If at all possible, use your wind turbine to equalize batteries in off-grid systems. You'd be amazed at how well this works. As a final note on the topic, be sure only to equalize flooded lead-acid batteries. A sealed battery, either gel cell batteries or absorbed glass matt sealed batteries, cannot be equalized! If you try to, you'll ruin the battery.

Reducing Battery Maintenance

Battery maintenance should take no more than 30 minutes a month. To reduce maintenance time, you can install sealed batteries. Another way to reduce time spent babying batteries is to replace factory battery caps with Hydrocaps (Figure 7.5). Hydrocaps capture much of the hydrogen and oxygen gases released by batteries when charging under normal operation. The gases are recombined in a small chamber in the cap filled with tiny beads coated with a platinum catalyst. Water formed in this reaction drips back into the batteries, reducing water losses by about 90 percent.

Another option is Water Miser caps. They capture moisture and acid mist escaping from batteries' fluid, reducing water loss by about 30 to 75 percent.

Yet another way to reduce maintenance is to install an automatic or semiautomatic battery filling system (Figure 7.6). Dan uses a manually operated Qwik-Fill battery watering system manufactured by Flow-Rite Controls in Grand Rapids, Michigan (sold

Fig. 7.5: *Hydrocaps. These simple devices help reduce battery watering by reducing water losses.*

JOE SCHWARTZ

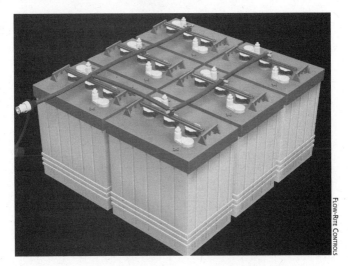

FLOW-RITE CONTROLS

Fig. 7.6: *Battery Filling System. Distilled water can be fed automatically to battery cells or manually pumped into them through plastic tubing. Both approaches save a lot of time and energy and help to keep battery fluid levels topped off to ensure battery longevity.*

Fig. 7.7: *Battery Filler Bottle. If you can access your batteries relatively easily, this filler bottle is one of the easiest and most economic means of adding distilled or deionized water to them.*

online through Jan Watercraft Products). He's found that this system works extremely well even after many years of service and has turned battery maintenance from a chore to a pleasure. Although they're a bit pricey, the systems quickly pay for themselves in reduced maintenance time and ease of operation. The convenience of quick battery watering overcomes the procrastination that leads to costly battery damage. A cheaper alternative is a half-gallon battery filler bottle (Figure 7.7).

Living with Batteries

Batteries work hard for those of us who live off-grid. They need to be properly installed and kept at the right temperature. Enclosures for flooded lead-acid batteries need to be vented. Batteries also need to be periodically filled with distilled water. You must also monitor their state of charge and recharge them quickly after deep discharging.

Generators in such systems need attention, too. If you install one in your system, you will need to periodically change oil and air filters. If you install a manually operated generator, you'll need to fire it up from time to time to raise the charge level or to equalize your batteries. You may also have to haul your generator in for an occasional repair.

In grid-connected systems with battery backup, you'll have much less to worry about. If you install sealed batteries, for example, you'll never need to check the fluid levels or fill batteries.

Batteries may seem complicated and difficult to get along with, but if you understand them, you can get lots of years of service from them. Break the rules and, well, you're going to pay for your carelessness.

CHAPTER 8

INVERTERS

The inverter is an indispensable component of virtually all electric-generating renewable energy systems. In this chapter, we'll discuss the types of inverters and the functions they provide. For a detailed look inside an inverter to see how it operates, you may want to pick up a copy of Dan's book, *Power from the Wind*.

Types of Inverters

Inverters come in three basic types: grid-connected, off-grid and grid-connected systems with battery backup.

Grid-Connected Inverters

Today, the vast majority of renewable energy systems — both wind and solar electric — are grid-connected. These systems require inverters that operate in sync with the utility grid and produce electricity that's identical to grid power.

Grid-connected inverters are also known as *utility-tie inverters*. They convert DC electricity from the controller in a wind system into AC electricity (Figure 8.1). Electricity then flows from the inverter to the breaker box and is then fed into active circuits, powering refrigerators, computers and the like. Surplus electricity is backfed onto the grid, running the electrical meter backward.

Grid-tied inverters produce electricity that matches the grid both in frequency and voltage. To do this, these inverters continuously

Fig. 8.1: *Grid-Connected Inverter. This inverter by Magnetek is designed for batteryless grid-connected PV systems.*

monitor the voltage and frequency of electricity on the utility lines. They adjust their output so it matches grid power. That way, electricity backfed from a wind-electric system onto utility lines is identical to the electricity that utilities are transmitting to their customers.

Grid-compatible inverters are equipped with *anti-islanding protection* — a feature that automatically disconnects the inverter from the grid in case of loss of grid power. That is, grid-connected inverters are programmed to shut down if the grid goes down. The inverter stays off until service is restored. This feature protects utility workers from electrical shock.

Grid-compatible inverters also shut down if there's an increase or decrease in either the frequency or voltage of grid power outside the inverter's acceptable limits (established by the utility companies). If either varies from the pre-programmed settings, the inverter turns off.

Grid-connected inverters also come with a *fault condition reset* — a sensor and a switch that turns the inverter on when the grid is back up or the inverter senses the proper voltage and/or frequency.

As noted in Chapter 5, the inverter shuts down, in part, because it requires grid connection to determine the frequency and voltage of the AC electricity it produces. Without the connection, the inverter can't operate. In most systems, the electrical output of the wind turbine is diverted to a dump load. In others, the controller shuts down the turbine.

In a grid-connected system with battery backup the inverters disconnect from the utility during outages, but continue to operate and can draw electricity from the battery bank to supply active circuits. Such systems, however, are typically designed to provide electricity only to essential circuits in a home or business, supplying the most critical loads.

Grid-connected inverters also frequently contain LCD displays that provide information on the input voltage (the voltage of the electricity from the turbine) and the output voltage (the voltage of the AC electricity the inverter produces and delivers to a home and the grid). They also display the current (amps) of the AC output.

Grid-connected inverters for wind systems are frequently sold with the wind turbine. Manufacturers specify the grid-tied inverters for their wind turbine because every turbine has a different output voltage range. One turbine may produce AC that ranges from 0 to 300 volts. Another may produce wild AC from 0 to 200 volts. Manufacturers select inverters with an input range that corresponds to the output voltage of the turbine.

Off-Grid Inverters

Rather than receiving electricity directly from the wind turbine, off-grid inverters typically receive their input from the battery bank. They convert the DC electricity from the battery bank into AC and boost the voltage to 120 or 240 volts. Off-grid inverters and inverters installed in grid-connected systems with battery backup also perform a number of other functions, described below. (We'll refer to these collectively as battery-based inverters.) If you're installing an off-grid system, be sure to read this carefully.

Battery-based inverters contain battery chargers. Battery chargers charge batteries from an external source — usually a gen-set in an off-grid system or the utility in a grid-connected system with battery backup. The battery charger in the inverter converts AC from the gen-set into DC electricity. It then feeds the DC electricity to the batteries.

In off-grid systems, battery charging gen-sets are used to restore battery charge after periods of deep discharge — if there's not enough wind or solar and wind energy. As noted in Chapter 7, this prolongs battery life and prevents irreparable damage to the plates. Battery chargers are also used during equalization.

High-quality battery-based inverters also contain high- and low-voltage disconnects. These features protect various components of a system, such as the batteries, appliances and electronics in a home or business. They also protect the inverters. To learn more about them, you may want to check out Dan's book, *Power from the Wind*.

Multifunction Inverters

Grid-connected systems with battery backup require multifunction inverters. They're also sometimes referred to as *multifunction* or, less commonly, *multimode* inverters (Figure 8.2).

Multifunction inverters contain features of grid-connected and off-grid inverters. Like a grid-connected inverter, they contain an anti-islanding feature that automatically disconnects the inverter from the grid in case of loss of grid power, over/under voltage or over/under frequency. They also contain fault condition reset — to power up an inverter when a problem with the utility grid is fixed. Like off-grid inverters, multifunction inverters contain battery chargers and high- and low-voltage disconnects.

If you are installing an off-grid system, you may want to consider installing a multifunction inverter in case you decide to connect to the grid in the future. Although multifunction inverters allow system flexibility, they are not always the most efficient inverters. That's because some portion of the electricity generated in such

Fig. 8.2: *This multifunction inverter from Xantrex is designed for grid-connected systems with battery backup.*

a system must be used to keep the batteries topped off. This may only require a few percent but over time, but a few percent add up. In systems with large battery banks, the electricity required to maintain them (to counter self-discharge) can be quite substantial. It is also worth noting that as batteries age, they become less efficient; more electricity is consumed to maintain the charge, which reduces the efficiency of the system. (And, as a rule, remember that older batteries lose charge faster than new batteries.)

If you want the security of battery backup in a grid-connected system, isolate and power only your most critical loads from the battery bank. This minimizes the size of the battery bank and reduces system losses and the cost of the system. Unless you suffer frequent or sustained utility outages, a batteryless grid-connected system usually makes more sense from economic and environmental perspectives.

Buying an Inverter

Most homes and small businesses require inverters in the 2,500 to 5,500-watt range. Which inverter should you select?

If you are installing a grid-connected wind system, the decision will be made for you by the manufacturer as noted earlier. If you are installing a battery-based system, you'll need a battery-charging wind turbine and an inverter that's compatible with batteries.

Most installers carry inverters they have a high degree of confidence in. Consequently, they will make a recommendation that fits your needs from their product line. Unfortunately, there are not many battery-based inverters available in North America.

System Voltage

When shopping for a battery-based inverter, you'll need to select one with an input voltage that corresponds to the battery voltage of your system. System voltage is the voltage of the electricity produced by the wind turbine. That is, the generators in these turbines are typically wired to produce 12-, 24- or 48-volt electricity. The batteries are wired similarly.

Because all components of an off-grid renewable energy system must operate at the same voltage, the inverter must match the source (wind turbine) and the batteries. If you are installing a 48-volt Bergey XL-R, you'll need a 48-volt battery-based inverter, and you must wire your battery bank for 48 volts. It is a good idea to talk with the wind turbine manufacturer to obtain their input on the best inverter.

Modified Square Wave vs. Sine Wave

The next inverter selection criterion is the output waveform. Battery-based inverters are available in *modified square wave* (often called modified sine wave) and *sine wave*. Grid-connected inverters are all sine wave so their output matches utility power. What does all this mean?

Waveform refers to the voltage of AC electricity as it changes over time (alternates). Modified square wave electricity is a crude approximation of the grid power voltage pattern. It works fairly well in many appliances and electrical devices in our homes. Although most all office and household electronic equipment and appliances can function on modified square wave electricity, they run less efficiently, producing less of what you want — i.e., light, water pumped, etc. — and more waste heat for a given energy input. When operated on modified square wave electricity, microwave ovens cook slower. Equipment and appliances that run warmer might last fewer years. Computers and other digital devices operate with more errors and crashes. Digital clocks don't maintain their settings as well. Modified square wave electricity may cause an annoying high-pitched buzz or a hum on TVs and stereos and may also produce annoying lines on TV sets. It can even damage sensitive electronic equipment. Some equipment, like modern washing machines, may not operate at all on modified sine wave electricity. The computer that controls these units won't run on it. Unless money is tight, get a sine wave battery-based inverter for an off-grid system.

Output Power, Surge Capacity and Efficiency

When selecting an inverter, even a grid-tied inverter, be sure to pay attention to continuous output, surge capacity and efficiency.

Continuous Output

Continuous output is the power an inverter can produce on a continuous basis. It is measured in watts, although some inverter spec sheets also list continuous output in amps (to convert, use the formula watts = amps x volts). For example, OutBack's sine wave inverter VFX3524 produces 3,500 watts of continuous power and is designed for use in 24-volt systems. The 35 in the model number stands for 3,500 watts. The 24 indicates it is designed for a 24-volt system.

To determine the continuous output you'll need, add up the wattages of the common appliances you think will be operating at once. Be reasonable, though. Typically, only two or three large loads operate simultaneously.

Surge Capacity

Electrical devices with motors, such as vacuum cleaners, washing machines and power tools, require a surge of power to start up. It typically lasts only a fraction of a second. Even though the power surge is brief, if an inverter can't provide the power, the motor won't start. Moreover, the stalled motor will draw excessive current and could overheat, unless it is protected by a thermal cutout. If not, it may burn out.

When shopping for an inverter, be sure to check out the surge capacity. All quality inverters are designed to permit a large surge of power over a short period, usually about five seconds. Surge power is listed on spec sheets in watts and/or amps.

Efficiency

Converting one form of energy to another results in a loss of energy. Efficiency is calculated by dividing the energy coming out by the energy going in. Fortunately, efficiency losses in inverters are quite low — usually only 5 to, at the most, 15 percent. It should be noted, however, that inverter efficiency varies with load. Generally, an inverter doesn't achieve its highest efficiency until output reaches 20 to 30 percent of its rated capacity. A 3,000-watt inverter, for instance, will be most efficient at outputs above 600 to 900 watts. At lower outputs, efficiency is dramatically reduced.

Noise and Other Considerations

Battery-based inverters are typically installed inside, close to the batteries to reduce line loss. Grid-tied inverters are almost always installed near the service entrance — where the utility service enters the house, which is near the breaker box. (Most inverter manufacturers like their equipment to be housed at room temperature.)

If you are planning on installing an inverter inside your home or office, be sure to check out the sound it produces. Inquire about this upfront. Ask to listen to the model you are considering in operation.

Some folks are concerned about the potential health effects of extremely low frequency electromagnetic waves emitted by inverters, electronic equipment and electrical wires. If you are concerned about this, install your inverter away from people. Avoid locations in which people will be spending a lot of time — for example, don't install the inverter on the other side of a wall from your bedroom or office.

Be sure to add ease of programming to the checklist of features to consider when purchasing an inverter. Find out in advance how easy it is to change settings. Spend some time with the manual.

Stackability

Finally, when buying a battery-based inverter, you may want to select one that can be "stacked" — connected to a second inverter of the same kind. Stacking permits homeowners to produce more electricity if demands increase over time. Two inverters can be wired in parallel, for example, to double the output (amps) of a battery-based wind system.

Stacking may also be needed to supply 240-volt AC electricity to operate appliances such as electric clothes dryers, electric stoves or central air conditioning. We recommend that you avoid such appliances, especially when installing an off-grid system. That's not because a wind or hybrid wind and solar system can't meet those needs, but rather because these appliances use lots of electricity and you'll need a very large and costly system to power them. Well-designed, energy-efficient homes can usually avoid using 240 VAC. An exception is a deep well pump, which may require 240-volt electricity. In most cases, high efficiency 120-volt AC pumps, or even DC pumps, perform admirably.

If you must have 240-volt AC electricity, purchase an inverter that can be wired in series to produce 240 VAC. Or you can purchase an inverter that produces 120- and 240-volt electricity. Or you can install a step-up transformer that converts 120-volt AC electricity from your inverter to 240-volt AC. Or you can simply install a dedicated 240-volt output inverter for that load.

Conclusion

A good inverter is key to the success of a wind system, so shop carefully. Size it appropriately. Be sure to consider future electrical needs. But don't forget that you can trim electrical consumption through conservation and efficiency. Efficiency is always cheaper than adding more capacity! When shopping, select the features you want and buy the best inverter you can afford. Although modified square wave inverters work for most applications, you will most likely be happier with a sine wave inverter. If you are installing a grid-connected system, you must install a sine wave inverter.

CHAPTER 9

FINAL CONSIDERATIONS

To those of us who love the idea of generating electricity from the wind, there are few things, if any, in the world more rewarding than watching a wind turbine's blades spin for the first time. Generating electricity from a clean, abundant and reliable source of energy is truly a thrill. The road from conception to installation of a turbine on a tower, however, can be long and arduous. The steps are summarized in Table 9.1. As you can see, so far we have only covered the first three steps. The rest of this chapter is devoted to the remaining steps.

Covenants and Neighbors' Concerns

If you want to install a wind system, the first step you must take is to check covenants — if you live in a covenanted community. Covenants govern many aspects of peoples' lives — from the color of paint they can use on their homes to the installation of privacy fences. They are typically found in subdivisions, even rural subdivisions. Covenants sometimes expressly prohibit renewable energy systems, such as solar electric systems. They may, however, be silent on wind systems, because these systems are rarely installed in neighborhoods.

Covenants from homeowner associations can create a huge obstacle to wind system installations. Even with permission from the zoning department and a permit from the local building department,

---

Table 9.1
Steps to Implement a Small Wind Energy Project

1. Measure your electrical consumption.

2. Assess your wind resource.

3. Select a turbine and tower.

4. Check zoning regulations and homeowner association regulations.

5. Check building permit and zoning requirements.

6. Check covenants.

7. Contact the local utility and sign utility interconnection agreement (for grid-connected systems).

8. Obtain building and electrical permits.

9. Order turbine, tower and balance of system.

10. Install system.

11. Commission — require installer to verify performance of the system.

12. Inspect and maintain the system on an annual basis.

JIM GREEN AND ROBERT PREUS

(discussed next) restrictive covenants in force in many neighborhoods can block an installation.

Unfortunately, courts have consistently upheld the legality of covenants. If you live in a covenanted community, contact the head of the homeowners' or neighborhood association to see what their rules are. Do this *before* you obtain a building permit.

Even in the absence of restrictive covenants, you should consider the needs and desires of your neighbors unless you live on a large piece of property in the country with few, if any, neighbors. Don't expect your neighbors to be as enthusiastic about a wind turbine as you are. In Chapter 1 we gave you some information to offer worried neighbors. For example, we gave you the facts about the terribly misinformed, but very common, concern over bird kills.

Be prepared to respond to all concerns without being defensive. You may want to show your neighbors pictures of the wind turbine and tower to allay their fears. If you are good at Photoshop, you can even take a photo of your home from their house and place a picture of the wind turbine and tower on your property so they can see what it will look like. If you are thinking about installing a grid-connected system, let them know that you'll be supplying part of *their* energy, too.

While you are at it, be sure to check out local sound ordinances, if any, to ensure you'll be in compliance. Let your neighbors know that you are concerned about protecting them from unwanted sound and are doing everything possible to safeguard them — for example, by mounting your turbine on a tall tower in a location on your property that minimizes sound beyond your property line.

Some companies offer Q & A sheets for residential wind machines. They address many common concerns and can be a great resource when warming your neighbors up to the idea of a wind turbine on your property. You might make copies to pass out in person to neighbors. (A typewritten summary of key points also works well.) Don't leave material in mailboxes. It's very likely not going to be read unless you hand it to your neighbors personally. Also, resist the temptation to overwhelm them with written material. It's best to talk to your neighbors in person so you can address their concerns directly — before they are blown out of proportion.

Giving neighbors advance notice, answering questions they have, and being responsive to their concerns is the best way to avoid misunderstandings and problems later.

Zoning and Permits

The next step when considering a wind system is to check on zoning regulations — legal restrictions imposed by local or state governments. These may include height restrictions or setbacks — how far a tower must be placed from property lines and utility lines.

Zoning Regulations

Wind turbines installations are typically subject to zoning laws. Although zoning regulations are in place in numerous cities and towns in North America, some rural areas have none. Homeowners are free to do whatever they want when building or installing a wind system — unless they're installing a grid-connected system (more on that shortly).

Before you purchase a system, check local zoning regulations to find out if it's legal to install a wind turbine, and, if so, what restrictions exist. A call to the local zoning and planning office will usually suffice. Building department officials will know, too. If a wind turbine is permitted, you may find that zoning regulations limit tower height to 35 feet, a useless height for those who want to generate electricity. To obtain permission to install a taller tower, you'll need to apply for a variance or a special use permit — legal permission that permits one to vary from zoning regulations. These are issued one property at a time.

Obtaining a variance or special use permit may take several months and can cost thousands of dollars. You begin by submitting a formal request to the zoning commission, then attending a hearing with them. You may need to hire an attorney to assist. If others have been granted a variance to install a tower over 35 feet in your jurisdiction, all the better. Legal precedent may make your job easier.

Variances are often granted to permit the construction of tall silos, radio and transmission towers, and cell phone towers. This information could help you obtain a variance as well. Be sure to explain the reasons why a tall tower is required, too. This strengthens your case.

Building and Electrical Permits

Most wind turbine installations also require a building permit. Building permits ensure that the project: (1) is safe, (2) is on your property within required setbacks, and (3) complies with local ordinances, including zoning, if any. They also ensure that your improvement is added to your property records for property tax assessment. Building permits cost from $50 to $6,000, depending on the jurisdiction. To determine if you need a building permit, contact the local building department. This may be a county government office or a city or town department.

Building permits are issued after a review of plans that include a site map, drawn to scale. It indicates property lines and other buildings on and near the site, including nearest neighbors. The map also shows topographic features, easements (if any), and the location and height of the proposed wind tower. Permit applications also indicate the kind of turbine and tower you'll be installing. If the zoning regulations require setbacks — placement of a tower a certain distance from property lines, streets or overhead utility lines — the site plan and building permit must indicate your compliance with them. Setbacks are required in many cases so that if a tower collapses, which is rare, it falls within the bounds of your property.

Although setbacks can affect the placement of a wind turbine, there are many exceptions made every year. If you can find exceptions in your area, precedent is in your favor. You can also speak with your neighbors about the setback. If they have no objection to the placement of a tower close to their property line, the building department may grant the permit.

Building permit applications often require drawings of the tower foundation and technical information on the wind turbine, including sound levels. Homeowners may rarely be required to submit an engineering analysis of the foundation and tower to demonstrate that both will be structurally sound and comply with local building codes. Engineering analyses are typically provided by the turbine and tower manufacturer and may be stamped by their own engineer.

Some municipalities require a stamp by a state-licensed engineer, which could cost $500, possibly more. Check requirements in advance. If this is the case, you may want to appeal the requirement. Companies manufacturing towers for wind systems must perform the engineering analysis to secure liability insurance.

Even if not required, it is always useful to submit drawings of the tower and the tower footings when requesting a building permit. This will assure building department officials that your project is well thought out and safe.

Electrical permits may be required to ensure systems are installed according to electrical code. Local officials use the National Electric Code (NEC) to govern the installation of all wiring and associated hardware such as inverters. Codes help protect against shocks and fires caused by electricity. Even if the building department does not require an electrical permit and inspections, the utility may want to inspect the system — if you are connecting to the grid — to be sure your system is safe.

Building department inspectors will visit your site at various stages to ensure that you are doing things correctly. An electrical inspector will also visit at least once to check wiring.

When applying for a building permit, be sure to make it clear that you are planning on installing a small, residential wind machine to offset your own personal electrical demand. Be sure to treat all code officials, including the folks at the front desk, with courtesy and respect. A qualified installer can handle a lot of this for you. Don't be intimidated by resistance from code officials in the office or at public hearings. You should expect to be treated in a reasonable and timely manner by everyone involved. Quite often, just letting a building department know that you have consulted with an attorney about your rights as well as your responsibilities will help make the process run more smoothly.

On another related issue, avoid the impulse to bypass the law — not obtain permits. This can create huge problems. Municipalities have the legal right to require a homeowner to remove his or her wind system if a building permit has not been secured. It's happened.

Obtaining a permit for a wind system may take several months, so apply several months in advance. Don't buy a wind energy system until your permit(s) have been granted.

Connecting to the Grid: Working with Your Local Utility

If you are installing a grid-connected system, you'll need to contact your local utility company to obtain a copy of their interconnection agreement. You'll need to assure them that the electricity you will be providing will be of the same quality as the power they supply to their customers. Utilities may have questions about what happens if the grid goes down and they need to send a lineman to work on it. They don't want their linemen getting shocked by a wind system that's operating while their system is down.

You won't need to agree on compensation for surplus electricity you will supply to the grid. The amount is stipulated by state law. All utilities in the United States are required by federal law (the Public Utility Regulatory Policy Act of 1978) to accept surplus electricity produced by their grid-connected customers. However, federal law only requires utilities to pay the "avoided cost" for monthly or annual surplus. Avoided cost is what it costs the utility to generate the power, which is usually one-fourth to one-third the amount they charge customers. If you are paying 8 cents a kilowatt-hour, the avoided cost may be as low as 2.5 cents. That's all the utility is required to pay.

Most states have enacted net metering laws that require utilities to reimburse residential customers at the avoided cost (Chapter 3). Many others do not pay anything for the net energy generation. A few pay the same rate they charge customers — the retail rate.

Before they sign an agreement, your utility may require you to submit simple electrical drawings that indicate where components will be located. Its two most important concerns, however, will be an assurance that you have liability insurance and that your equipment includes some sort of system for automatically disconnecting the wind turbine from the grid should there be a power outage. This is a standard feature on grid-connected inverters. Even so, the utility may require a demonstration or an inspection.

Utilities may also require you to install a visible, lockable AC disconnect — a manually operated switch that allows utility workers to disconnect your wind system from outside your home in case they need to work on the lines. This is required even though grid-connect inverters automatically terminate the flow of electricity to the grid when they detect a drop in line voltage or a change in frequency. Do not fight your utility on this redundancy. While you may be in the right, keep in mind that the utility is a lot bigger than you are, with much more legal expertise and practically unlimited financial resources to fight your installation. In a case of irreconcilable differences, your best bet is to contact the public utility commission for help. If you go into the process with the proper knowledge, however, it is unlikely ever to come to that.

Insurance Requirements

To protect against damage to your system and damage caused by your system, a property owner should secure insurance. Two types of insurance are required: property damage, which protects against damage *to* the wind system and liability insurance to protect you against damage *caused by* the system.

Insuring Against Property Damage

For homeowners, the most cost-effective way to insure a wind system against damage is under an existing homeowner's insurance policy. Businesses can cover a system under their property insurance.

When installing a system, contact your insurance company to determine if your current coverage is sufficient. If not, boost the coverage to cover replacement, including materials and labor. Wind systems are insured as "appurtenant structures" on homeowners' policies. An appurtenant structure is any uninhabited structure on your property not physically attached to your home. Examples include unattached garages, sheds and satellite dishes. Appurtenant structures are assessed and charged at a lower rate than occupied structures. Insurance companies usually base premiums for appurtenant structures on the total cost of materials plus the labor to build the structure.

It is best, although a bit more expensive, to insure a system at its full replacement cost — not a depreciated value. Wind energy systems can easily last two decades or more. Systems should have insurance coverage that includes damage to the system itself from "acts of nature," plus possible options for theft, vandalism or flooding.

Insuring a wind system is relatively inexpensive. While home insurance coverage should cover appurtenant structures, added insurance may be required. It can be purchased for an additional premium.

Liability Coverage

Liability insurance is also part of homeowner's or business owner's insurance policy. It protects against possible damage to others caused by a system. For example, it would cover you if neighbors claimed their electronic equipment was damaged because of your grid-connected system. (Your system would somehow have to be able to magically send power onto the line to damage electronic equipment in a neighbor's home. This, of course, is not possible.) It also covers personal injury or death of employees due to electrical shock from a system when working on a utility line during a power outage. Even though the likelihood of these events is nil, because of the automatic disconnect feature built into inverters, utilities may insist on this coverage. Don't resist them.

Liability coverage is relatively inexpensive. In most places, liability coverage for homes runs from $100,000 to $300,000. In most areas, increasing liability coverage to $500,000 may add an additional $10 to an annual premium. Extending coverage to $1 million may add $35 to $40 more to the annual premium. The utility will likely dictate the level of liability insurance that it requires as a condition for interconnection. If the amount required by the utility seems unreasonable to you, consider appealing to your state's public utility commission. A utility may simply make a recommendation. Dan's rural electric co-op at his educational center, The Evergreen Institute, recommends $1 million liability.

Buying a Wind Energy System

If wind energy seems like a good financial, social or recreational pursuit for you, we recommend hiring a competent, experienced professional wind site assessor and a professional installer. A local supplier/installer with experience and knowledge will supply all of the equipment, be certain that it is compatible, help obtain permits and install the system. He or she will test the system to be sure it is operating satisfactorily and will be there to answer questions and to address problems you may experience.

We recommend that you find an installer who also provides routine maintenance and one who stands solidly by his or her work. Most wind turbines are guaranteed for five years, though that's changing, but you want to be sure the installer will respond promptly and knows what he or she is doing. Look for someone who's been in the business for a while — the longer, the better — and who has already installed a lot of similar systems.

As in any major home project, ask for references, and call them. Visit wind installations, if possible, and contact the local office of the Better Business Bureau. Get everything in writing. Sign a contract. Be sure the installer has insurance to protect employees during installation. Don't pay for the entire installation up front. Be sure you are on the site when the work is done.

You can also purchase equipment from a local or online supplier, and put it up yourself with or without their guidance. We don't recommend this route unless you are handy and you have attended a couple of workshops. Raising a tower is risky business. Wiring is fraught with difficulties. Connecting to the electrical grid is a job for professionals.

Another option is to buy equipment and supplies from a local supplier or an online source and sponsor a workshop on your property to take care of the installation. Organizations like Dan's, The Evergreen Institute Center for Renewable Energy and Green Building, Solar Energy International, and the Midwest Renewable Energy Association are often looking for wind energy installations in different parts of the country. They bring in wind energy experts who teach a one- or two-week installation workshop. Although you may not save any money on the installation, workshop installations can be very satisfying. Be sure your insurance will cover volunteer workers on your site.

Maintaining a Wind-Electric System

Wind turbines work long and hard in a severe environment. "Wind turbines are machines. They have moving parts," notes John

Hippensteel owner of Lake Michigan Wind and Sun. "That means they require maintenance."

Maintenance takes time and money.

Ignoring routine maintenance will surely reduce performance and will very likely lead to even more costly maintenance in the future — or even failure. Don't be lured by exaggerated claims of maintenance-free operation made by wind turbine manufacturers or installers. No wind generator is "maintenance free" — unless you leave it in its box.

This section briefly discusses maintenance of wind energy systems (except for batteries, covered in Chapter 7). A more detailed account is offered in Dan's book, *Power from the Wind*. This material is intended to apprise readers of a vital, though sometimes overlooked, part of living with a wind-electric system. It is not meant to discourage you from pursuing wind energy, but rather to drive home the importance of maintenance requirements and let you know what you're getting into. You should embark on this venture with your eyes wide open. Remember: you can hire a professional to inspect and maintain your system.

What's Required

We recommend twice-a-year inspections for all small wind systems. The first inspection should be in the spring once the weather warms, but before thunderstorm season. Thunderstorms are likely the most violent weather your turbine will encounter, and you want to be sure it's in good shape before Nature's onslaught. The second inspection should occur in the fall to be sure your turbine is ready for winter. The last thing you will want to do is climb a tower in a 30 mile-per-hour wind at -30° F to take care of something that you could have been fixed on a warm fall day.

Inspections should be performed on windless days. This makes climbing the tower or lowering the tower easier, safer, and more comfortable. Wind turbines on climbable towers should be shut down to prevent injury. Never perform inspection and maintenance when the blades are spinning. After you've climbed the tower, tie the blades to the tower so they can't spin, just in case the winds start up while you're working on the turbine. Also, be sure to brake a wind turbine on a tilt-up tower before lowering it.

What constitutes inspection and maintenance will vary with the particular wind turbine and type of tower. Installation manuals that come with the turbine and/or tower usually provide detailed instructions. Be sure to read them carefully. Before performing your semi-annual maintenance, call the manufacturer for any maintenance updates. It's a good idea to check for updates every year if you have installed a relatively new turbine — one that's been released within the last five years. For turbines that have been in the field for long periods updates are very likely going to be rare.

Before you climb a tower or lower a turbine to the ground in the case of tilt-up towers, be sure to listen for unusual sounds — for example, clunking, banging or grinding. They indicate a bad alternator or bearings. Check for vibrations. Unusual sounds from a wind turbine should be noted any time it is operating.

You should also test the electrical production of the turbine. Detailed instructions on how to do this should be in the installation manual. Contact the manufacturer if you have any questions on test procedures. Be sure to talk to engineers, not sales personnel. Any unusual sound or low output should be taken as an indication that a detailed inspection is necessary.

Inspecting a wind system requires a careful examination of the turbine, including the blades, tail, tail boom and internal components; the tower; and guy cables, if any, and connections to anchors. You'll also need to inspect the electrical wires and connections. Loose nuts in the tower and turbine will need to be tightened according to the manufacturer's recommendations. Cracked nuts and bolts must be replaced.

You also need to check for rust, cracked welds, and cracks in the tower and the turbine. Cracks or rust must be attended to immediately. Cracked welds indicate that something is seriously wrong. Lower the tower and fix them ASAP.

Be sure to examine the furling mechanism, including the cable that is used to manually furl the tail, if present. When you have reached the base, be sure the furling winch, if any, is in good shape and operating correctly. Also, check the connections on ground rods. Be sure that all contact surfaces are clean and free of oxidation. Inspect other electrical components such as surge arrestors for damage and replace them, if necessary. If your system is equipped with a dynamic brake (the mechanism that shorts the alternator to stop the turbine), be sure to check the switch.

If overspeed protection is provided by a vertical furling mechanism or a blade pitch mechanism, be sure to inspect the bolts, pins and springs for damage. Tighten loose bolts per manufacturers' recommendations and grease as needed.

When climbing a tower, be sure to wear a safety harness (preferably a full-body harness). When climbing a tower or working on a platform, be sure you are securely tied in or clipped in at all times. Secure all tools and extra parts so they don't fall and injure people or pets below you. We discussed safe climbing in Chapter 6. A more detailed account is provided in *Power from the Wind*.

Be sure to read the entire inspection and maintenance protocol *before* ascending the tower to be certain you have all the tools and parts you need. Make a checklist of tools and inspection and maintenance duties on the front and back of a note card. Carry it with you to the top of the tower. You might want to put a duplicate copy in your pocket in case a breeze comes and sends your list on a cross-country flight.

If your turbine is mounted on a tilt-up tower, you will need to lower it to inspect the turbine. You'll need to inspect the guy cables, their attachment to the anchors, the tower, and electrical wire that carries electricity from the tower to the balance of system.

Routine inspection and maintenance of a wind turbine and tower is an insurance policy. Like taking your car in for oil changes and other scheduled maintenance, it ensures longer life and can save a lot of money over the long haul. Tightening a loose nut is a lot cheaper than replacing a part, or worse, an entire wind turbine that's been damaged when a nut comes off. Proper maintenance can help the system last for 20 to 30 years — helping you get the most for your investment.

If you install a battery bank, you'll need to inspect and maintain the batteries as well. Off-grid battery banks typically require a lot more care than batteries installed in grid-connected systems. In contrast, inverters are generally fairly maintenance free. The only attention they require is when they break down, which is rare.

If you're not up to the task or can never seem to get around to chores, you may want to hire a professional to inspect your turbine, tower and other components every year. If that's not an option, you may want to think about a less maintenance-intensive renewable energy system, like a PV system.

Parting Thoughts

When you started reading this book, you may have simply wanted to determine if a wind system was suitable for your home or business. Perhaps you were sure you wanted to install a wind system, but didn't know how to proceed. We hope that you now have a clear understanding of what is involved.

If you have come to realize that your dreams for a wind system were unrealistic, you can pursue other dreams — and be glad that you did not spend a lot of money on a wind system that would not have met your expectations. If, however, you now know that a wind system is right for you, the next step is to investigate further, find a dealer/installer and start harnessing the wind.

Our job is over. We've helped you get up to speed on wind energy systems. You've obtained a good, solid working knowledge, with good theoretical as well as practical information, that puts you in a good position to move forward.

May the wind be as much an ally to you as it has been to us!

Resource Guide

Books

Bartmann, Dan and Dan Fink. *Homebrew Wind Power: Build Your Own Wind Turbine: A Hands-On Guide to Harnessing the Wind*. Buckville Publications, LLC, 2008. A clear and comprehensive guide to building a quiet, efficient, reliable and affordable wind turbine.

Chiras, Dan. *Power from the Wind*. New Society Publishers, 2009. The unabridged version of this book.

Chiras, Dan. *The Homeowner's Guide to Renewable Energy*. New Society Publishers, 2006. Contains information on residential wind energy and solar electric systems.

Gipe, Paul. *Wind Power: Renewable Energy for Home, Farm, and Business*. Chelsea Green, 2004. Somewhat technical introduction to small and large (commercial) wind generators.

Gipe, Paul. *Wind Energy Basics: A Guide to Small and Micro Wind Systems*. Chelsea Green, 1999. A brief and somewhat technical introduction to wind energy for newcomers.

Piggott, Hugh. *Wind Power Workshop*. Center for Alternative Technology, 1997. A guide for those who want to build their own wind turbines and towers by Europe's small wind expert.

Magazines

BackHome P.O. Box 70, Hendersonville, NC 28793. Tel: (800) 992-2546. Website: BackHomeMagazine.com. Publishes articles on renewable energy and many other subjects for those interested in creating a more self-sufficient lifestyle.

Backwoods Home Magazine P.O. Box 712, Gold Beach, OR 97444. Tel: (800) 835-2418. Website: backwoodshome.com. Publishes articles on all aspects of self-reliant living, including renewable energy strategies.

Home Power P.O. Box 520, Ashland, OR 97520. Tel: (800) 707-6585. Website: homepower.com. Publishes numerous extremely valuable how-to and general articles on renewable energy, including solar hot water, PVs, wind energy, microhydroelectric and occasionally an article or two on passive solar heating and cooling. This magazine is a goldmine of information, an absolute must for anyone interested in learning more. It also contains important product reviews and ads for companies and professional installers. They also sell CDs containing back issues.

Mother Earth News 1503 SW 42nd St., Topeka, KS 66609. Website: motherearthnews.com. One of my favorite magazines. Usually publishes a very useful article in each issue on some aspect of renewable energy.

Solar Today ASES, 2400 Central Ave., Suite G-1, Boulder, CO 80301. Tel: (303) 443-3130. Website: solartoday.org. This magazine, published by the American Solar Energy Society, contains lots of good information on passive solar, solar thermal, photovoltaics, hydrogen and other topics, but not much how-to information. Also lists names of engineers, builders, and installers and lists workshops and conferences. Mick Sagrillo writes a monthly column for them.

Organizations

American Wind Energy Association, 122C Street, NW, Suite 380, Washington, D.C. 20001. Tel: (202) 383-2500. Website: ogc.apc.org/awea. This organization sponsors an annual conference on wind energy. Their website has a list of relevant publications, their online newsletter, frequently asked questions, news releases and links to companies and organizations.

British Wind Energy Association, 26 Spring Street, London W2 1JA. Tel: 0171 402 7102. Website: bwea.com. Actively promotes wind energy in Great Britain. Check out their website for fact sheets, answers to frequently asked questions, links and a directory of companies.

Centre for Alternative Technology, Machynlleth, Powys SY20 9AZ, UK. Tel: 01654 703409. Website: cat.org.uk. This educational group in the United Kingdom offers workshops on alternative energy, including wind, solar and micro hydroelectric.

Centre for Renewable Energy and Sustainable Technologies, Holywell Park, Loughborough University, Loughborough, Leicestershire, LE11 3TU. Website: lboro.ac.uk/crest. Active in research and education in renewable energy.

European Wind Energy Association, Rue du Trone 26, B-1000, Brussels, Belgium. Tel: +32 2 546 1940. Website: ewea.org. Promotes wind energy in Europe. The organization publishes the *European Wind Energy Association Magazine*. Their website contains information on wind energy in Europe and offers a list of publications and links to other sites.

Energy Efficient Building Association, 490 Concordia Ave., P.O. Box 22307, Eagen, MN 55122. Tel: (651) 268-7585. Website: eeba.org. Offers conferences, workshops, publications and an online bookstore.

National Wind Technology Center of The National Renewable Energy Laboratory, 1617 Cole Blvd., Golden, CO 80401-3393. Tel: (303) 275-3000. Website: nrel.gov/wind. Their website provides a great deal of information on wind energy, including a wind resource database.

Solar Energy International, P.O. Box 715, Carbondale, CO 81623. Tel: (970) 963-8855. Website: solarenergy.org. Offers a wide range of workshops on solar energy, wind energy and natural building.

Solar Living Institute, P.O. Box 836, Hopland, CA 95449. Tel: (707) 744-2017. Website: solarliving.org. A nonprofit organization that offers frequent hands-on workshops on solar energy and many other topics. Be sure to tour their facility if you are in the neighborhood.

US Department of Energy and Environmental Protection Agency's ENERGY STAR program. Tel: (888) 782-7937.
Website: energystar.gov.

US Department of Energy's Energy Efficiency and Renewable Energy.
Website: eere.energy.gov.

Small Wind Turbine Manufacturers

Abundant Renewable Energy, 22700 NE Mountain Top Road, Newberg, OR 97132. Tel: (503) 538-8298. Website: abundantre.com.

Aerostar, Inc., PO Box 52, Westport Point, MA 02791. Tel: (508) 636-5200. Website: aerostarwind.com.

Bergey Windpower, 2200 Industrial Blvd., Norman, OK 73069. Tel: (405) 364-4212. Website: bergey.com.

Bornay Wind Turbines, P.I. RIU, Cno. Del Campanar, 03420 Catalla (Alicante) Spain. Tel: 965 560-025; 966 543 077.
Website: bornay.com.

Flowtrack Australia, End of Turntable Falls Road, Nimbin, Australia. Tel: 0266 891431. Website: flowtrack.com.au.

Iskra Wind Turbines, Ltd., The Innovation Centre, Epinal Way, Loughborough LE11 3EH, United Kingdom. Tel: 0845 8380588. Website: iskrawind.com.

Jacobs Wind Turbines, Wind Turbine Industries Corp., 16801 Industrial Circle S.E., Prior Lake, Minnesota 55272. Tel: (952) 447-6064. Website: windturbine.net.

Kestrel Wind Turbines, P.O. Box 3191, North End, Port Elizabeth 6056, South Africa. Tel: +27 (0)41 401 2645.
Website: kestrelwind.co.za.

Marlec Engineering Co. Ltd., Rutland House, Tevithick Road, Corby, Northants NN17 5XY, United Kingdom. Tel: +44 (0)1535 201588. Website: marlec.co.uk

Proven Energy Ltd., Wardhead Park, Stewarton, Ayrshire, KA3 5LH, Scotland, UK. Tel: 0044 (0) 1560 485 570.
Website: provenenergy.co.uk.

Southwest Windpower, 1801 W. Route 66, Flagstaff, AZ 86001. Tel: (928) 779-9463. Website: windenergy.com.

West Wind, J.A. Graham Group, 3 Carmavy Road, Crumlin, Co. Antrim, BT29 4TF, Northern Ireland. Tel: 0044 (0) 28 9445 2437. Website: westwindturbines.co.uk.

Wind Turbine Tower Kits

Abundant Renewable Energy, 22700 NE Mountain Top Road, Newberg, OR 97132. Tel: (503) 538-8298.
Website: abundantre.com.

Bergey Windpower, 2200 Industrial Blvd., Norman, OK 73069. Tel: (405) 364-4212. Website: bergey.com.

IDC Solar, Wind & Water, P.O. Box 630, Chino Valley, AZ 86323. Tel: (928) 636-9864. Website: idcsolar.com.

Independent Power Systems, 1501 Lee Hill Road #19, Boulder, CO 80304. Tel: (303) 443-0115. Website: solarips.com.

Lake Michigan Wind and Sun, 1015 County Road U, Sturgeon Bay, WI 54235. Tel: (970) 743-0456. Website: windandsun.com.

Otherpower.com, 2606 W. Vine Drive, Fort Collins, CO 80521. Tel: (877) 944-6247. Website: otherpower.com.

Southwest Windpower, 1801 W. Route 66, Flagstaff, AZ 86001. Tel: (928) 779-9463. Website: windenergy.com.

Index

About the Author

Dan Chiras is an internationally acclaimed author who has published over 24 books, including *The Homeowner's Guide to Renewable Energy* and *Green Home Improvement*. He is a certified wind site assessor and has installed several residential wind systems. Dan is director of The Evergreen Institute's Center for Renewable Energy and Green Building (www.evergreeninstitute.org) in east-central Missouri where he teaches workshops on small wind energy systems, solar electricity, passive solar design and green building. Dan also has an active consulting business, Sustainable Systems Design (www.danchiras.com) and has consulted on numerous projects in North America and Central America in the past ten years.

Dan lives in a passive solar home powered by wind and solar electricity in Evergreen, Colorado.

Dan Chiras.

If you have enjoyed *Wind Power Basics* you might also enjoy other

BOOKS TO BUILD A NEW SOCIETY

Our books provide positive solutions for people who want to make a difference. We specialize in:

**Sustainable Living • Green Building • Peak Oil
Renewable Energy • Environment & Economy
Natural Building & Appropriate Technology
Progressive Leadership • Resistance and Community
Educational and Parenting Resources**

New Society Publishers

ENVIRONMENTAL BENEFITS STATEMENT

New Society Publishers has chosen to produce this book on Enviro 100, recycled paper made with **100% post consumer waste**, processed chlorine free, and old growth free.

For every 5,000 books printed, New Society saves the following resources:[1]

15	Trees
1,357	Pounds of Solid Waste
1,494	Gallons of Water
1,948	Kilowatt Hours of Electricity
2,468	Pounds of Greenhouse Gases
11	Pounds of HAPs, VOCs, and AOX Combined
4	Cubic Yards of Landfill Space

---

[1]Environmental benefits are calculated based on research done by the Environmental Defense Fund and other members of the Paper Task Force who study the environmental impacts of the paper industry.

For a full list of NSP's titles, please call **1-800-567-6772**
or check out our website at: **www.newsociety.com**

NEW SOCIETY PUBLISHERS